Antarctica
Global Science from a Frozen Continent

Antarctica is the coldest and driest continent on Earth – a place for adventure and a key area for global science. Research conducted in this extreme environment has received increasing international attention in recent years due to concerns over destruction of the ozone layer above it and the problems of global warming and rising sea levels. Data collected in the Antarctic now informs a wide range of scientific fields. A record of the globe's climate is locked up in its deep snow and ice while, as part of the early supercontinent Gondwana, its rocks have much to teach us about the geological history of the Earth. A diversity of unique plants and animals abound in Antarctic waters and the clear skies overhead allow astronomers to probe the outer reaches of the Universe.

Governed internationally since 1959, the Antarctic is also an object lesson in collaboration between nations. This dramatically illustrated new book brings together an international group of leading Antarctic scientists to explain why the Antarctic is so central to understanding the history and potential fate of our planet. It introduces the beauty of the world's greatest wilderness, its remarkable attributes, and the global importance of the international science done there.

Spanning topics from marine biology to space science, this book is an accessible overview for anyone interested in the Antarctic and its science and governance. It provides a valuable summary for those involved in polar management and development of new research programmes, and is an inspiration for the next generation of Antarctic researchers.

Professor David Walton began work in 1967 with the British Antarctic Survey (BAS). He is now an Emeritus Fellow at BAS, publishing papers and books on many aspects of Antarctica. Professor Walton represented the international Antarctic scientific community at Antarctic Treaty Meetings for 14 years and was awarded the first SCAR medal for International Scientific Coordination. He was also awarded a Polar Medal by the Queen. The author of more than 100 scientific papers, and more than 250 reviews, popular articles and reports, he has also written and edited several books and has been the editor in chief of the international journal *Antarctic Science* for the last 25 years.

'I am privileged to have visited Antarctica twice, as Chief Executive of the UK Natural Environment Research Council. It was wonderful to have the work being done at the British Antarctic Survey's base at Rothera explained to me personally by some of the world's leading Antarctic scientists. When I retired I hugely missed these "personal tutorials", but here is the answer! A book packed with the most up-to-date information about the history, geology, biology, the changing climate, human physiology, oceanography, and the space-science and politics of Antarctica, edited and written by world experts but accessible to anybody interested in this wonderful frozen continent. And above all, it explains why Antarctica is so fundamental to the scientific understanding of the future of Planet Earth.'

- **Professor Sir John Lawton CBE FRS**,
last chair of the Royal Commission on Environment Pollution

'This is an excellent review of key multidisciplinary collaborative research and geopolitics in Antarctica involving more than 30 countries, addressing global issues in climate, oceans, biodiversity, solar system, tourism and more. Of importance to contemporary society, it is a valued compendium.'

- **Dr Alan K. Cooper**,
Consulting Professor, Stanford University and recipient of the second SCAR medal for International Scientific Coordination

'Antarctica is a conundrum. It is distant, yet it will shape our children's future; it is mysterious, yet a treasure trove for science; it is the focus of calculated geopolitical interest, yet the exemplar of "world governance". In this well-presented and readable book, the world's leading experts on Antarctic science showcase why the uninhabited seventh continent is central to the present and future of human interests and wellbeing.'

- **Professor Chris Rapley CBE**,
University College London; former Director of British Antarctic Survey and former President of SCAR

Community Learning & Libraries
Cymuned Ddysgu a Llyfrgelloedd

This item should be returned or renewed by the last date stamped below.

To renew visit:

www.newport.gov.uk/libraries

ANTARCTICA

Global Science from a Frozen Continent

Edited by

DAVID W. H. WALTON

British Antarctic Survey

CAMBRIDGE
UNIVERSITY PRESS

CAMBRIDGE UNIVERSITY PRESS
Cambridge, New York, Melbourne, Madrid, Cape Town,
Singapore, São Paulo, Delhi, Mexico City

Cambridge University Press
The Edinburgh Building, Cambridge CB2 8RU, UK

Published in the United States of America by
Cambridge University Press, New York

www.cambridge.org
Information on this title: www.cambridge.org/9781107003927

First published 2013

Printed and bound by Grafos SA, Arte sobre papel, Barcelona, Spain

A catalogue record for this publication is available from the British Library

Library of Congress Cataloguing in Publication data

Antarctica : global science from a frozen continent / edited by David W. H. Walton.
     p. cm.
  Includes index.
  ISBN 978-1-107-00392-7 (Hardback)
1. Antarctica–Discovery and exploration.   I. Walton, D. W. H.
  G860.A5575 2013
  559.89–dc23

                                                      2012026684

ISBN 978-1-107-00392-7 Hardback

Contents

List of contributors   *vii*

Introduction   *xi*
David W. H. Walton

1 **Discovering the unknown continent**   *1*
David W. H. Walton

2 **A keystone in a changing world**   *35*
Bryan Storey

3 **Ice with everything**   *67*
Valérie Masson-Delmotte

4 **Climate of extremes**   *102*
John J. Cassano

5 **Stormy and icy seas**   *137*
Eberhard Fahrbach

6 **Life in a cold environment**   *161*
Peter Convey, Angelika Brandt and Steve Nicol

7 **Space science research from Antarctica**   *211*
Louis J. Lanzerotti and Allan T. Weatherwax

8 **Living and working in the cold**   *229*
Lou Sanson

9 **Scientists together in the cold**   *253*
Colin P. Summerhayes

10 **Managing the frozen commons**   *273*
Olav Orheim

11 **Antarctica: a global change perspective**   *301*
Alan Rodger

Appendix A   **Visiting Antarctica**   *325*
Appendix B   **Further reading**   *328*

Acknowledgements   *333*
Index   *334*

Contributors

Professor Angelika Brandt is Professor of Zoology at the University of Hamburg and Director of the Zoological Museum. Her research focuses on systematics, evolution, ecology, biogeography and biodiversity of peracarid crustaceans in the deep sea, especially in the polar regions. Named a National Geographic Adventurer of the Year in 2007, Angelika has been a driving force in the international Census of Marine Life and is the author of over 100 papers.

Associate Professor John Cassano teaches and researches in the Department of Atmospheric and Oceanic Sciences, University of Colorado, and is a fellow of the Cooperative Institute for Research in Environmental Sciences at the University of Colorado. He uses observations from autonomous observing systems and computer models of the atmosphere to study the weather and climate of the polar regions. He has spent ten field seasons in Antarctica since 1994 and is the author of three books and 60 scientific papers.

Professor Peter Convey, a terrestrial ecologist with British Antarctic Survey for 24 years, has authored over 200 publications on polar biology. Active in the development of international Antarctic science through the Scientific Committee on Antarctic Research, he is an Honorary Lecturer at the University of Birmingham and a Visiting Professor at the University of Malaya. He is an advisory editor for the journals *Global Change Biology* and *BMC Ecology*.

Dr Eberhard Fahrbach studied physics in Heidelberg and physical oceanography at Kiel University, initially working on wave structure off Sierra Leone. His PhD in 1984 was on the heat budget of the equatorial Atlantic. In 1986 he moved to the Alfred Wegener Institute for Polar and Marine Research where he has studied the role of the Antarctic and Arctic Oceans in the climate system, leading many cruises to the polar oceans. He was awarded the Georg Wüst Prize in 2007 for his outstanding contribution to oceanography.

Professor Louis J. Lanzerotti, now retired from the Bell Laboratories Lucent Technologies, is currently Distinguished Research Professor of Physics, New Jersey Institute of Technology, Newark. He has extensive research experience in the Antarctic and with spacecraft instruments, all concerned with studies of the Earth's space environment. He is a Fellow of several professional societies and a member of the National Academy of Engineering and the International Academy of Astronautics.

Dr Valérie Masson-Delmotte is a senior scientist and head of a research group at Laboratoire des Sciences du Climat et l'Environnement (LSCE), France, using natural archives to investigate past climate dynamics. She analyses ice cores and tree rings to quantify past changes in both climate and the water cycle on various time scales (centuries, glacial–interglacial cycles and abrupt events), together with climate models to understand these changes. She has published over 120 papers and books on climate change.

Dr Stephen Nicol worked at the Australian Antarctic Division for 24 years, first heading the krill research, then as a Program Leader for Southern Ocean Ecosystem Studies. A member of the Australian delegation to the Commission for the Conservation of Antarctic Marine Living Resources (CCAMLR) from 1987–2010, much of his research has been directed towards marine conservation. Now an Adjunct Professor at the Institute for Marine and Antarctic Studies, University of Tasmania, he has published widely and has visited Antarctica several times. He was awarded the Australian Antarctic Medal in 2011.

Dr Olav Orheim is a glaciologist with over 80 research publications. Head of Antarctic Research at the Norwegian Polar Institute 1972–93 he served as Director 1993–2005. A member of Norwegian delegations to the Antarctic Treaty Consultative Meetings for three decades, he was elected as the first Chair of the Committee on Environmental Protection, 1998–2002, and Chair of the Legal and Institutional Working Group, 2005–09. He is presently Chair of the Board of several Norwegian institutions, including the polar ship *Fram* Museum in Oslo.

After a degree in electronic engineering from the University of Manchester, Professor **Alan Rodger** wintered in the Antarctic in the early 1970s as an electronics technician with the British Antarctic Survey. His career evolved into a research scientist focussing on upper atmospheric physics. He now leads the interdisciplinary research programmes at BAS and provides advice to the government on climate change in the polar regions. He is Visiting Professor at Manchester.

Lou Sanson is Chief Executive of Antarctica New Zealand and responsible for the New Zealand Antarctic Programme. He has served on the executive of the Council of Managers of National Antarctic Programs. His career has been in environmental management with oversight of significant environmental projects such as the establishment of

New Zealand's Subantarctic World Heritage Area, Stewart Island National Park, the ANDRILL sedimentary drilling project in Antarctica and New Zealand's role in the International Polar Year.

Bryan Storey became Professor of Antarctic Studies and the first Director of Gateway Antarctica, the Centre for Antarctic Studies and Research at the University of Canterbury, in 2000. Prior to this he spent 24 years with the British Antarctic Survey leading and developing research programmes on the geological evolution of Gondwana. He was awarded a Polar Medal in 1987 for his outstanding achievements in Antarctic research. He has authored over 120 scientific papers.

Dr Colin Summerhayes, a marine geochemist, is an Emeritus Associate at the Scott Polar Research Institute (SPRI) of Cambridge University. He has worked in academia, government and industry (BP and EXXON), while living in the UK, New Zealand, the USA, South Africa and France. He has published 8 books and 230 research papers and other articles. He is the immediate Past President of the Society for Underwater Technology, and a Vice President of the Geological Society of London.

Allan Weatherwax is Professor of Physics and Dean of the School of Science at Siena College, in Loudonville, New York. He directs ground-based experiments in Antarctica, Canada and Greenland that explore the Earth's upper atmosphere and co-directs the satellite mission Firefly that will explore gamma rays produced by lightning discharges. He serves on numerous national and international committees and is a member of the Polar Research Board of the National Academy of Sciences.

Introduction

The polar regions have long exerted a fascination for politicians, scientists and the public in many countries. The early concerns were a mixture of scientific investigations (especially for magnetism) and the drive to explore and annex unknown lands, a feature of imperial powers of the nineteenth and twentieth centuries. Thus, the United Kingdom, along with Germany, France and Sweden, all made important contributions in the early twentieth century to those pioneering explorers from Britain, Imperial Russia, France and the United States who all claim a part in the original discovery of the continent in the eighteenth and early nineteenth centuries.

Despite the heroic national expeditions led by Scott, Shackleton, Amundsen, Bruce, Mawson, Drygalski, Filchner, Nordenskjöld and Charcot early last century, the Antarctic was seen as a backwater by most scientists, ignored as having little of significance to contribute to major scientific paradigms, whilst its remoteness and climate provided almost insurmountable obstacles to access and work there.

How dramatically this has changed is now familiar to many people throughout the world as during the last 60 years the continent has been the subject of increasing international research of the most sophisticated sort. With the initial stimulus of the International Geophysical Year to highlight the Antarctic as one of the two great 'Unknowns' (the other being space), those countries with scientists active there have grown from 12 to over 30, a remarkable international legal instrument has developed to manage not only the continent but the surrounding Southern Ocean, and the science has gone from initial survey and list making to addressing global problems such as the impacts of climate change on world sea level, marine biodiversity and its origins, the history of world climate over the past 1 million years, the relationships between the Sun and the physics of the Earth's atmosphere, and the origins of the southern hemisphere continents. Antarctic science is no longer a backwater but a major contributor to our understanding of the way in which the world works – the holistic study we now call Earth System Science – and discoveries there make major news stories throughout the world and documentaries on television.

Even with this growth in science on and around the continent the number of people able to actually visit Antarctica has been very limited – perhaps only around 250 000 in total as part of national expeditions over a period of more than 100 years. So an equally important change has been the enormous growth since 1970 in the number of people who have been able to visit the continent as tourists. This must now exceed 300 000 and forms an international cohort of those for whom the Antarctic is no longer a visual abstraction but a real place with very particular qualities.

This general growth in interest and importance for Antarctic science means that there is now a need to explain much more clearly what we know about Antarctica, how its position, its history and its special qualities contribute to our understanding of global problems, and what the future might hold for this international continent governed by a consensus representing over 65% of the global population. Whilst this volume does not cover absolutely every area of Antarctic science, it does provide not only an explanation of all the key areas but also, through its international authorship, epitomises that element of collaboration which underpins everything we do in this frozen continent.

No longer a remote continent at the end of the world of little significance, Antarctica in the twenty-first century is now recognised as a key part of the global jigsaw, a driver for the models of our likely future climate and a continuing test bed for the success of international political collaboration and cooperation. Its future matters to everyone.

1 Discovering the unknown continent

DAVID W. H. WALTON

We had discovered a land of so extensive a coastline and attaining such an altitude as to justify the appellation of a Great New Southern Continent. **James Clark Ross, 1847**

Finding the poles

There have always been those who have wanted to see over the horizon, to cross the lakes and seas, and climb the mountains. Explorers led the great migrations as humans spread across the globe, successively colonising all the continents except Antarctica, and slowly but steadily taming the wild places for their own uses. The lure of the unknown has persisted for millennia and continues to drive those today who search for new knowledge and new frontiers. Their stories have fascinated the public for centuries and no more so than those from the exploration of the polar regions.

As long ago as 1531 geographers had concluded that for the Earth to exist the two hemispheres of the globe must be balanced. If there was significant land in the north towards the pole there must be a similar mass in the south. It was but a small step then to imagine a large and populous southern continent, probably extending into temperate latitudes. Spain and Portugal had already found riches to plunder and native people to convert and subjugate in South America so perhaps, thought the geographers, Antarctica would yield a similar prize. The maps they produced in the late sixteenth century contained much cartographic speculation, which provided a stimulus for sailors from many European countries to set out across the world in search of fame and fortune. Finding Antarctica and beginning its scientific investigation would prove a lengthy business and one that would turn out to be truly international.

Antarctica: Global Science from a Frozen Continent, ed. David W. H. Walton. Published by Cambridge University Press. © Cambridge University Press 2013.

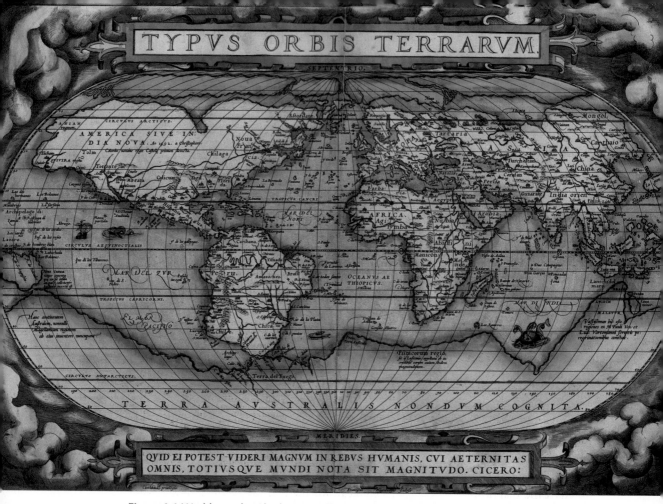

Figure 1.1 World map by Abraham Ortelius published in 1570.

As shipping expanded and explorers sailed into the unknown it became essential to improve navigation and position fixing. Magnetic compasses had been in use since the twelfth century and by the sixteenth century it had been realised that there was a difference between the North Magnetic Pole and the geographic North Pole. Britain, as the global power, decided it was essential to improve knowledge of the magnetic fields so that compasses could be used more accurately by allowing for changes in magnetic variation. Edmond Halley, a leading astronomer and Fellow of the Royal Society, was commissioned to sail to the South Atlantic in 1698 aboard the *Paramour* and measure the strength of the magnetic fields. Reporting on icebergs at 52°S, only 200 miles west of South Georgia, his voyages could be said to mark the beginning of the exploration of the Southern Ocean and the Antarctic.

It was not only the British Government who felt responsibilities for world leadership. The Royal Society in London, established as the world's first permanent scientific society in 1660, was keen to build on Halley's experience and improve

Figure 1.2 Jasperware medallion of Captain James Cook, by Josiah Wedgewood and Thomas Bentley, 1779 (Credit: British Museum).

astronomical knowledge, and had agreed with the Admiralty that an expedition should go to the Pacific Ocean to observe the rare transit of Venus across the Sun in 1769. Appointing Captain James Cook to lead this and two subsequent expeditions, the Royal Society instructed him to sail as far south as possible and search for the southern continent proposed by Alexander Dalrymple, who had argued that 'such a large continent must have many millions of inhabitants and that under the sovereignty of Britain it would generate more than enough trade to replace that lost with the ungrateful subjects in the American colonies'. Thus was set the scene for the voyages that would provide the truth about Antarctica and lay the historical foundations for all our present activities there.

Departing in August 1768, this first voyage, as well as Cook's second circumnavigation in 1773–75, laid to rest this preposterous hypothesis and began to chart the real limits of the Southern Ocean for the first time. Others had already ventured around the world, led by the Portuguese explorer Ferdinand Magellan in 1519–22, and followed by Sir Francis Drake, William Dampier, George Anson, John Byron and Louis de Bougainville, but nobody until Cook made a determined effort to assess the southern continent.

His assessment for the future was bleak:

The risk one runs in exploreing a coast in these unknown and Icy Seas, is so very great, that I can be bold to say, that no man will ever venture farther than I have done and that the lands which may lie to the South will never be explored

And although his reports of fur seals stimulated uncontrolled exploitation by British and American sealers, the failure to find an Antarctic land of riches dissuaded others from exploring the region for almost 50 years. After all, there was so much else of the world still to explore and Africa and Asia seized the interest of both governments and companies. It was not until James Clark Ross sailed to the Ross Sea in 1839–43 aboard the *Erebus* and *Terror* that any major progress with exploration was made by government expeditions.

Whilst the activities of the sealers ranged across all the sub-Antarctic islands and down south to the Antarctic Peninsula most of them had little interest in the region

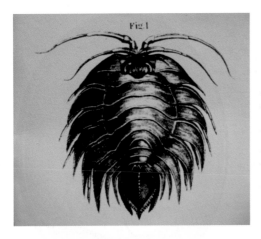

Figure 1.3 One of the first Antarctic benthic organisms discovered, *Glyptonotus antarcticus*, described and illustrated by James Eights in 1853.

other than killing as many animals as possible. They kept their sealing grounds secret to limit competition and the ships' logs and other written materials provide very limited evidence about where they went and when. This has led to continuing argument over who first saw the continent and who landed there first. It does seem likely, however, that Captain John Davis from the *Cecilia*, tender to the American sealer *Huron*, was the first to land on the Antarctic Peninsula, probably around Hughes Bay (64° 20′S 61° 15′W) on 7 February 1821.

James Weddell was a sealer who reached as far south as 74° 15′ in the sea that now bears his name, a record that lasted for almost 90 years. This remarkable discovery of open water so far south was certain to be doubted so Weddell had the accuracy of the log witnessed by his Chief Officer and two seamen. Unusually for a sailor, he was interested in science, well read and with a good command of English. His published account of his voyage aboard the *Jane* (1822–24) contains many interesting scientific observations on ice, penguins, seals and geology and it was he who first collected *Leptonychotes weddelli*, the Weddell seal.

Some, like the American James Eights, used the sealers to visit the Antarctic. In 1829–31 he visited Staten Island, Tierra del Fuego and the South Shetland Islands. Collecting specimens of rocks, lichens and marine animals he was the first to describe the geology of the South Shetlands, provide lists of the birds, seals and whales that he saw and describe three new species of benthic animals. His collection of the grass he found was identified by the Royal Botanic Gardens, Kew, and was the first scientific record of any flowering plant in Antarctica. Despite his important scientific achievements – his scientific papers form the start of what we might consider as the 'professional' scientific output from Antarctica – he did not go on the later United States Exploring Expedition led by Charles Wilkes and his work remained almost unknown for nearly 100 years.

In the intervening period a major Russian expedition, led by the Estonian Thaddeus Fabian von Bellinghausen, was sent in 1819 by Tsar Alexander I to explore towards the South Pole. This was imperial exploration meant to directly rival the activities of the British Navy and make sure that Russia was seen as a great power. His discoveries of Peter I Island, the first land to be sighted within the Antarctic Circle, and the South Sandwich Islands, where he recorded volcanic activity for the first time in the Antarctic region, were valuable additions to existing knowledge. He may also have been the first to sight the Antarctic continent, if the edge of the continental ice shelf that he reported on 28 January 1820 is considered as part of the continent. Interestingly, Bellinghausen speculated that the South Sandwich Islands and South Georgia might lie on a submarine mountain chain, a prescient recognition of what we now know as the Scotia Ridge.

In 1828 the Royal Navy sent HMS *Chanticleer* down the Atlantic to the South Shetland Islands to make gravity measurements, using a pendulum as well as magnetic and meteorological observations. The ship reached Deception Island in 1829 and stayed there 2 months, allowing the surgeon, William Webster, to make extensive natural history collections, including the first algae to be brought back from the Antarctic.

There were other expeditions during this early period of the nineteenth century. The first to be conceived was the United States Exploring Expedition led by Lieutenant Charles Wilkes. It took a long time to persuade Congress of the value of the expedition, not least because one of its chief supporters was an enthusiast for the 'hollow earth' hypothesis. Finally leaving in 1838, the expedition was equipped with four inadequate vessels and failed to attract the necessary scientific expertise, with only one of the scientists that did sail (Titian Ramsay Peale) actually involved in the Antarctic legs of the cruise. Wilkes and the officers made meteorological and magnetic measurements whilst Peale collected some biological specimens. When Wilkes returned after an acrimonious voyage he was court-martialled for a range of petty offences largely dreamed up by his officers. He was acquitted and later rose to be an admiral in the US Navy. He put great energy into publishing the scientific results, very few of which were Antarctic. However, it was this expedition that collected and described the keystone marine species *Euphausia superba*, or Antarctic krill as it is commonly known, and the main food of birds, seals and whales.

Meanwhile the French were much better organised with an expedition led by Captain Jules-Sébastien-César Dumont d'Urville. Dumont d'Urville proposed the expedition in January 1837 and sailed in September of the same year in the two corvettes *Astrolabe* and *Zélée*. A gifted linguist and scholar Dumont d'Urville had already circumnavigated the world twice and was desperate to continue his studies in Pacific ethnology. He certainly had no plans for investigating the polar regions but the King insisted he sail towards the South Pole, apparently for the glory of France. Since the British, Russians and Americans were busy searching for the southern continent France could not be left behind. He initially tried to follow Weddell's tracks but it was a bad ice year and he could not get far south so sailed west to the South Orkney Islands where the ships became trapped in the pack ice for a terrifying period of 5 days. Finally breaking free, he sailed south-west to the South

'PEACOCK' IN CONTACT WITH ICEBERG.

Figure 1.4 Engraving by Albert Thomas Agate of the USS *Peacock* in pack ice, one of the ships in the United States Exploring Expedition commanded by Charles Wilkes.

Shetland Islands and down to the tip of the Antarctic Peninsula. Here he found new land and islands to map and name, claiming the territory for France as Louis-Philippe Land. By now it was March and with many sailors suffering from scurvy Dumont d'Urville felt he had done enough, sailing north to Chile and then on west. After spending 2 years in the Pacific he determined to claim the South Magnetic Pole for France on the way back. His party landed on 21 January 1840 on a small islet they called Pointe Geologie, as close as they could get to the Magnetic Pole, raising the Tricolour and toasting their success at finally reaching the continent. Dumont d'Urville claimed the coast for France, naming it Adélie Land after his wife. In the official scientific reports of the expedition Antarctica plays only a minor part, but the expedition did discover the crabeater seal and several new benthic organisms.

James Clark Ross had spent 15 summers and 8 winters in the Arctic before he sailed south to explore the Antarctic on 5 October 1839. A superb navigator and seaman, his earlier polar work had awakened an enthusiasm for science, and his natural history contributions had been recognised by his election to the Linnean Society in 1823. His expedition was also fortunate to be staffed with several full-time

Figure 1.5 Commemoration of James Clark Ross in a stamp for British Antarctic Territory.

naturalists of whom the most outstanding proved to be Joseph Dalton Hooker, later to be Director of Kew Gardens. The scientific equipment he was provided with was exceptional and a library of scientific books was carried. The Admiralty provided him with a precise range of tasks focused on magnetism and exploration, whilst the Royal Society's instructions occupied a small book dealing with every possible science opportunity they could think of. Provided with two vessels, *Erebus* and *Terror*, whose design had already been tested in the Arctic, Ross ensured that the provisioning was adequate to stop scurvy breaking out. This was one of the most successful early Antarctic expeditions, making major advances in a wide range of fields. Certainly one of the most important was the provision of definitive charts of magnetic declination, dip and intensity replacing those developed by Wilkes and Dumont d'Urville. His discovery of the enormous Ross Ice Shelf and its identification as the source of tabular icebergs was an important step forward in glaciology and it was this expedition which discovered the Ross seal, deep in the pack ice. Ross was the first to make deep-sea soundings but with inadequate equipment some of his data proved to be erroneous. However, Hooker's *Flora Antarctica*, which encompassed all the plants he found both in the Antarctic and on the sub-Antarctic islands, remains a major reference work to this day. The *Flora* contains a detailed section on phytoplankton which probably constitutes the first recognition of their true significance in the world's oceans and the starting point for what is nowadays a major branch of oceanography. Surprisingly, neither Ross nor Hooker were keen advocates for further visits to the region and interest in magnetism had dwindled so, except for the Transit of Venus Expeditions by the UK, France and Germany to the sub-Antarctic islands, there was little official interest in the south for almost 50 years.

Matthew Maury, now thought of as the father of oceanography, was appointed Director of the US Naval Observatory in 1844. His compilations of data on winds and currents proved invaluable to shipping all over the world and his textbook *The Physical Geography of the Sea*, published in 1855, showed his recognition that the high southern latitudes held a crucial key to the weather of the southern

Figure 1.6 Karl Weyprecht, originator of the idea of International Polar Years.

hemisphere. His proposals for a US expedition to investigate this fell on deaf ears so in 1860, addressing the Royal Geographical Society in London, he suggested it was a task for the British.

It took a while for the idea to take hold but when the Royal Society decided to promote the importance of oceanographic cruises to the British Government the Admiralty became interested. Work in the North Atlantic from HMS *Lightening* in 1868 and HMS *Porcupine* in 1870 by Charles Wyville Thomson and William Carpenter had not only disproved the dogma that there was no life in the deep oceans but had also provided data to support Alexander von Humbolt's earlier idea that it was cold water from the Antarctic that drove the deep currents in the Atlantic towards the equator. Eventually, HMS *Challenger* was made available for oceanographic work and sailed in December 1872 on its epic voyage, returning in May 1876. The extensive results from the cruise appeared in 50 volumes, providing a wealth of data on oceanography and marine biology from around the world. The Antarctic leg was not long but the data showed the connections between Antarctic waters and the rest of the oceans for the first time.

Karl Weyprecht, the Austrian co-leader of the Austro-Hungarian Exploring Expedition to the Arctic, came up with the idea of collaborative science in the polar regions and his ideas eventually led to the first International Polar Year in 1882. The only Antarctic component was a German expedition to South Georgia, strongly supported by Dr Georg von Neumayer, then Director of the Deutsche Seewarte in Hamburg.

The Heroic Age

Scientists from several countries had begun to get interested in the Antarctic towards the end of the nineteenth century. The enthusiasm culminated in a major address by von Neumayer at the 6th International Geographical Congress in London in 1895 on the importance of Antarctic exploration and the need to undertake it immediately. Strongly supported by John Murray and Clements Markham, the UK began developing its plans to explore the continent and reach the South Pole. But they were not alone, as delegates from several other countries had

Figure 1.7 Sketches of whales by Emil Racovitza on board the *Belgica* during the Belgian Antarctic Expedition 1897–99.

returned home inspired by the possibilities and set about raising the necessary funds. The first to get underway was the expedition in the *Belgica* under the leadership of Lieutenant Adrien de Gerlache with Romanian, Polish and American members as well as Belgians. With little support from the Belgian government, no scientific programme and poor leadership, it is amazing what they managed to achieve. The ship became frozen in on the west coast of the Peninsula and the expedition was thus the first to over-winter in the Antarctic. One man died, two others went mad and more would have done so if the doctor Frederick Cook had not taken charge of morale. The scientific results included the first winter meteorological measurements, descriptions of the physical geography and petrology of the Peninsula and the first collections of terrestrial invertebrates.

Meanwhile, a Norwegian had organised an unusual British expedition aboard the *Southern Cross* to go to the Ross Sea in 1898. All except three members of the expedition were Norwegian but the principal backer, Sir George Newnes, insisted that it sailed under the British flag. The leader, Carsten Borchgrevink, had visited the Antarctic earlier with Henryk Bull's Norwegian expedition in

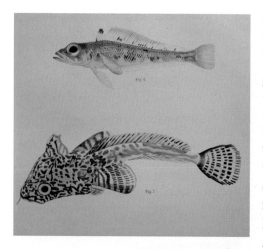

Figure 1.8 New species of Southern Ocean fish collected on the Swedish Antarctic Expedition, painted by Carl Skottsberg and described by Einar Lönnberg. They are, from the top, *Notothenia larseni* and *Artedidraco skottsbergi*.

1894 aboard the *Antarctic*, sent to investigate whaling opportunities. That expedition was the first to land on the continent itself (as distinct from the Peninsula) and the first to find vegetation within the Antarctic Circle. Borchgrevink's party also landed at Cape Adare and spent the winter there in a stone-built hut. The expedition was the first to overwinter on land, the first to make a sledge journey on the Ross Ice Shelf, whilst their meteorological records for the whole year were important in giving the first detailed picture of the continental climate.

The international scramble for exploring the continent suddenly increased. First the British organised their National Antarctic Expedition (1901–04) led by Captain Robert Scott on board the *Discovery*, whilst the Germans funded the first German Antarctic Expedition (1901–03) on the *Gauss* with Erich von Drygalski as its leader. Both of these expeditions carried considerable numbers of scientists and their reports can be seen as providing the beginnings of modern Antarctic science.

Meanwhile over in Sweden, Otto Nordenskjöld wanted to build on his experiences in Greenland and the Yukon and put together a privately funded expedition which sailed from Gothenburg in October 1901. The loss of his ship *Antarctic* in the ice and the rescue of all of his men against the odds made this one of the more exciting expeditions. Their collections of fossils identified Mt Flora and the islands around Hope Bay as one of the most important palaeontological sites in the Antarctic.

With one British expedition already in the Antarctic it might seem strange that a second private expedition should come from the United Kingdom. Organised as a declaration of Scottish nationalism, Dr William Bruce took the Scottish National Antarctic Expedition south on the *Scotia* in 1902 and wintered at Laurie Island in the South Orkney Islands. At the conclusion of the expedition, Bruce offered the meteorological station he had established to the Argentine Government – who accepted with alacrity and have staffed it even since. This has since provided the longest running meteorological record for anywhere in the Antarctic.

Figure 1.9 Omond House, Laurie Island, February 1904, with the Scottish National Antarctic Expedition on the left (Martin, Harvey-Pirie, Ross, Cuthburtson, Mossman and Smith) and the Argentine meteorologists on the right (Acuña, Szmula and Valette). (Courtesy National Archives, previously published in *The Log of the Scotia*, Edinburgh University Press, 1992)

With all these countries already active French nationalism insisted they must not be left out. Jean Charcot, a doctor with private means and an interest in sailing, persuaded the French Government to help with funding the French Antarctic Expedition aboard the *Français*. Charcot explored the west coast of the Antarctic Peninsula from 1903 to 1905. This whetted his appetite for polar work and he returned again in the *Pourquoi-Pas?* (1908–10) to extend his mapping of the coastline. In true French fashion Charcot recognised the importance of food to an expedition. Not for them the rigours of hard tack. Instead he laid in significant quantities of wine, rum and brandy as well as French delicacies, and hired a cook who could provide regular fresh bread and croissants on Sundays!

Still to come were the British expedition led by Ernest Shackleton that made the first ascent of Mt Erebus and failed to reach the South Pole, the Norwegian expedition lead by Roald Amundsen that successfully made it to the Pole, the second Scott expedition that made it to the Pole but died on the return journey, the first Japanese expedition lead by Lieutenant Nobu Shirase that achieved little, the second German expedition lead by Wilhelm Filchner aboard the *Deutschland*,

Figure 1.10 Chicks of two Antarctic birds – snowy petrel to the left and sheathbill to the right – painted by H. Goodchild on the Scottish National Antarctic Expedition 1902–04.

the first Australian expedition led by Douglas Mawson and the second Shackleton expedition whose attempt to cross the continent was defeated by the sinking of the *Endurance* in the Weddell Sea. Much has been written about all of these heroic attempts to map, chart and describe this inhospitable continent and the privations they suffered. It is hard in the twenty-first century to imagine how they managed to achieve so much with inadequate clothing, transport and equipment.

Most of the Heroic Age expeditions aimed to contribute to science as well as explore the Antarctic. The thrust of the British effort was to be first to reach the South Pole but, despite that, both of R. F. Scott's expeditions made substantial contributions to a wide variety of scientific fields. Although Roald Amundsen's Norwegian party beat Scott's party to the South Pole, his triumph was eclipsed by the tragic death of Scott and his men on the way back to the coast. Amundsen's planning and techniques were far superior to Scott's and for him it was enough simply to be first. He refused to be encumbered by the requirements of science and whilst history reminds us that he won the race, it is the legacy of the science from the British expedition that has contributed so much to modern Antarctic science.

Figure 1.11 Illustration of the meteorological screens from the *South Polar Times*, the midwinter book written and illustrated by Captain R. F. Scott's British National Antarctic Expedition 1901–04.

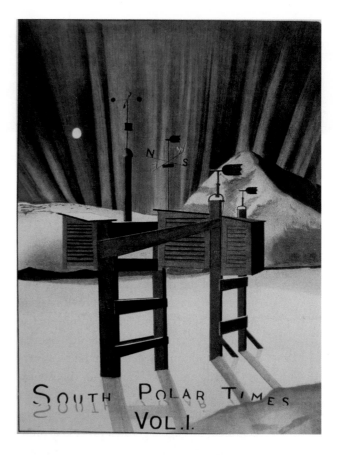

Enlightened imperialism

Britain had played a major role in exploring Antarctica and had claimed major areas of it for the Crown, an imperial approach that few other countries had followed. After Captain C. A. Larsen established the first whaling station on South Georgia, the British Government recognised the importance of legal controls over what would become the most profitable whaling area in the world. The Letters Patent issued in 1908 claimed the Falkland Islands Dependencies and provided control over all the shore-based whaling from South Georgia to the Peninsula. This allowed the then Governor of the Falklands, William Allardyce, to introduce regulations to protect female whales with calves and restrict the number of whaling stations and catchers. This must have been the first conservation initiative in the Antarctic and paved the way for the later establishment of the Discovery Investigations at South Georgia whose scientific reports have proved invaluable to all who have since worked on the Southern Ocean.

Figure 1.12 Grytviken, the first whaling station established on South Georgia photographed around 1920. (Credit: R. I. Lewis Smith)

After the flush of 'Heroic Age' expeditions World War I severely limited international interest in the Antarctic. It was during the war that Ernest Shackleton's ship the *Endurance* sank in the Weddell Sea at the start of his expedition to cross the continent. He and his companions made the heroic boat journey in the *James Caird* from Elephant Island to South Georgia in order to find help to rescue the crew, providing one of the most adventurous tales of Antarctic survival and an adventure that several groups have tried to repeat since. Interestingly, it was at South Georgia that Shackleton finally died in 1923. Sailing south on the *Quest*, he had a heart attack whilst the ship was anchored at Grytviken and was buried on the island.

National interest was, however, still alive in some quarters. The British Government had decided in 1919–20 to aim for control over the whole continent through a gradual process of annexation. In 1923 the Ross Dependency was claimed followed in 1933 by Australian Antarctic Territory, with the BANZARE Expedition of 1929–31 led by Douglas Mawson as preparation for this. Indeed, during a meeting in 1926 the Cabinet discussed the possibility of declaring the whole of Antarctica British, but decided against it as by then the French had laid claim to Adélie Land and the United States had begun its explorations with Lincoln Ellsworth and Richard Byrd. Norway had already expressed concerns that British claims encroached on areas discovered by Norwegians but did nothing about this until 1939 when, hearing that the German *Schwabenland* expedition had set off for 'their area' of the continent, the Norwegian Government finally claimed Dronning Maud Land.

In 1924 the United States Government applied the so-called 'Hughes doctrine' to Antarctic claims, which required effective occupation of an area before the United States would recognise the claim. The British activities did not meet this strict definition and thus the United States refused to recognise British claims as legal, a source of considerable irritation between London and Washington. The vacillations of the American government over whether or not it would make its own claim continued for many decades, despite claims being made in its name by Byrd.

Figure 1.13 The hut at Cape Royds, Ross Island, was built in 1908 by Ernest Shackleton's British Antarctic Expedition 1907–09. This photo shows how it appeared in 1956. Today it is under the management of New Zealand's Antarctic Heritage Trust. (Credit: Dick Prescott, NSF)

In fact, the State Department continued to secretly amass evidence on which to make an American claim right up until the mid-1950s whilst all the time suggesting that the claims of the UK, France, Norway, Argentina, Chile, Australia and New Zealand were unacceptable.

Argentina had disputed British title to the Falkland Islands and to South Georgia from the late nineteenth century. During World War II both Chile and Argentina decided to pursue rival claims to the area covered by the Falkland Islands Dependencies and thus initiated a long-running dispute with military overtones. Argentina's creation of a National Antarctic Commission in 1939 was prompted by the Norwegian claim to a sector. They despatched expeditions to Grahamland and the South Shetland Islands in 1942 and 1943, hoisting an Argentine flag over Deception Island. Their formal claim in 1943 was announced on the basis that the Argentine Antarctic sector had been an integral part of Argentina since independence in 1810 and this was now reinforced by polar activities as well as geological continuity. Chile's earlier 1940 claim was also based on a combination of

Figure 1.14 Tabarin nightclub poster from Paris before World War II.

historical, geographical and geological arguments. At a meeting on 28 January 1943, the War Cabinet decided to mount a secret naval expedition to the Antarctic, code named Operation Tabarin after a Paris nightclub. The real reason was to strengthen the legal grounds for opposing the Argentine and Chilean claims to sovereignty but this would have been unacceptable to the United States. Argentina's wartime links with Germany as well as the actions of the German Navy in the South Atlantic provided a convincing cover story that Operation Tabarin was needed to forestall German threats to British convoys and monitor warships around Cape Horn. The public objectives were to establish several stations to watch for German raiders, but the actual orders were to establish more or less permanent occupation and to remove any marks of Chilean or Argentine sovereignty. Building stations initially at Deception Island and at Hope Bay Operation Tabarin had a quiet war with no Germans sighted. Transferred to the Colonial Office in 1945 and renamed the Falkland Islands Dependencies Survey (FIDS) the primary objective of the stations remained political, to strengthen the British title to the area, and a programme of mapping was started in support of this. But the stations were now being staffed with scientists measuring the climate, surveying the geology and researching the flora and fauna.

The United States first major Antarctic foray of the twentieth century was led by Richard Byrd in 1928 with what was then the best equipped expedition to go to the Antarctic. He returned again in 1934 and yet again in 1939, establishing several temporary stations and pioneering the routine use of aircraft in Antarctic exploration. Meanwhile, the United States had certainly noticed the expansionist activities of Britain, Argentina and Chile and became concerned that it might be seen as uninterested in the continent, despite the US Antarctic Service Expedition in 1939–41, which had established two stations – Little America III and East Base on Stonington Island – that the US Government thought at that time might be used to mark the edges of its potential claim. Accordingly the Antarctic Developments Project in the US Navy organised Operation Highjump in 1946–47, by far the largest and most wide-ranging Antarctic expedition ever. With 13 ships, 33 aircraft and nearly 5000 personnel, it dwarfed every other country and, whilst apparently only there to test equipment and conduct aerial photomapping, it achieved its primary political objectives of displaying American power in a new theatre. There had also been the privately organised Ronne Antarctic Research Expedition to Stonington Island 1946–48 as further evidence of renewed American involvement.

Argentina had not given up its claim, and with the election of President Juan Peron in 1946, Antarctica reached a new level of importance in Argentine politics. Argentina sent an expedition of seven ships in 1946–47 to establish new stations to add to 'Orcadas', the station in the South Orkney Islands inherited from the Scottish National Antarctic Expedition in 1904. Chile was also mobilising and in 1947 established its first station 'Arturo Pratt' on Greenwich Island. The following year the President of Chile, Gonzalez Videla, visited the South Shetland Islands to inaugurate the second station 'Bernardo O'Higgins' on the Trinity Peninsula, the first ever visit by a head of state to the continent.

Relations continued to deteriorate with a clash at Hope Bay where Argentine soldiers fired over the heads of British scientists, and Britain took to deporting Argentine personnel from Deception Island as well as dismantling Argentine and Chilean stations. Protest notes were delivered from warships and claimant signs thrown into the sea. The risk of escalation was clearly recognised, as was the cost of keeping all this military hardware in the Antarctic. A tripartite agreement in 1949 between the three countries limited the deployment of warships south of 60° S, and this agreement was renewed annually for another decade until superseded by the Antarctic Treaty.

Figure 1.15 A Crown Land sign at Signy Island, South Orkney Islands, which has now disappeared. (Credit: R. Burton)

Meanwhile the first planned international scientific expedition to the continent was getting underway in Dronning Maud Land. Led by John Giaever the Norwegian–British–Swedish expedition also had Australian, Canadian, South African and Finnish participants. The three main objectives were glaciology, meteorology and geology as well as aerial surveying. Although during the expedition three men died, a great deal of new science was achieved. Perhaps the most important result was the seismic traverse inland in which it proved possible to map the underlying rock, the pilot study for the detailed radio echo sounding work carried out later by many nations to provide the present bedrock map of the Antarctic continent itself.

Elsewhere in Antarctica other countries were also establishing research stations. The Australian National Antarctic Research Expeditions was established in 1947 for purely political reasons, to counter any threats to Australian Antarctic Territory by Norway or the United States. Establishing stations first on Macquarie and Heard Islands, the Australians began to build their first continental station – Mawson – in 1954. The French were also actively beginning expeditions to Antarctica in 1948 and building their first station – Port Martin – in Terre Adélie in 1950.

Figure 1.16 Charcoal sketch of huskies made during a visit to British stations on the Antarctic Peninsula in the late 1970s. (Credit: David Smith)

International Geophysical Year: the turning point in Antarctic affairs

The negotiations for the International Geophysical Year (IGY) brought a truce between all three parties and Argentina, Britain and Chile joined with nine other nations to make the IGY a remarkable success in both scientific and political terms.

The IGY was conceived by a group of physicists on 5 April 1950. British physicist Sidney Chapman had stopped by in Washington to talk to James van Allen, a leading American space scientist. Over dinner with other physicists at van Allen's house, talk turned to the polar regions and what had been achieved in earlier International Polar Years. Recognising the opportunities of new technology, such as rockets and radar, the scientists became enthused with the idea of a new Polar Year, and that led to the proposal to the International Council of Scientific Unions (ICSU) in 1951, which in turn resulted in the IGY. Of great significance for success was the endorsement of IGY by President Eisenhower, and the appointment of

Figure 1.17 Vivian Fuchs, left, and Admiral George Dufek (US Navy) standing alongside Fuch's Snow-Cat a few minutes after the crossing party arrived at the South Pole on 20 January 1958. (Credit: NSF)

Laurence Gould, former chief scientist on the first Byrd expedition and a powerful figure in American academic circles, to lead it in the United States.

This scheme for international research in the polar regions needs to be seen in the context of the time. The Soviet Union and the United States were competing superpowers in an increasingly militarised world, there were Antarctic territorial disputes between Argentina, Chile and the UK, and the world at large was still recovering economically from World War II. It was clear to everyone that the Antarctic research proposed would be very expensive yet, despite the political and economic obstacles, it took place with all the nations working together and in agreement. It stimulated the space race by suggesting that earth orbit satellites would be very useful, an objective that persuaded the Soviets to rush to launch Sputnik 1 on 4 October 1957, soon to be followed by American satellites. It also precipitated a long-held ambition of Vivian Fuchs, then the Director of the Falkland Islands Dependencies Survey. He wanted to succeed in crossing the continent where Shackleton had failed. Backed by grants from the British, New Zealand, Australian and South African governments, as well as by donations from companies and individuals, the joint British–New Zealand Trans-Antarctic Expedition, lead by Vivian Fuchs and Edmund Hillary, finally crossed the surface of the continent in a traverse of 3472 km in 99 days.

Twelve nations signed up to take part in the Antarctic component of IGY. Argentina, Australia, Chile, France, New Zealand and the UK already had established stations but in some cases decided to build new ones for the IGY. Japan, Belgium, Norway, South Africa, the Soviet Union and the United States had to build their facilities from scratch.

For all the new aspirants the logistics were a serious problem but for the United States, which elected to build at the South Pole, and for the Soviet Union, at the Pole of Inaccessibility, the difficulties were enormous. In both cases military might was the order of the day with the navy, the army and the air force involved along with the US Coastguard who supplied the icebreaking capacity. This had

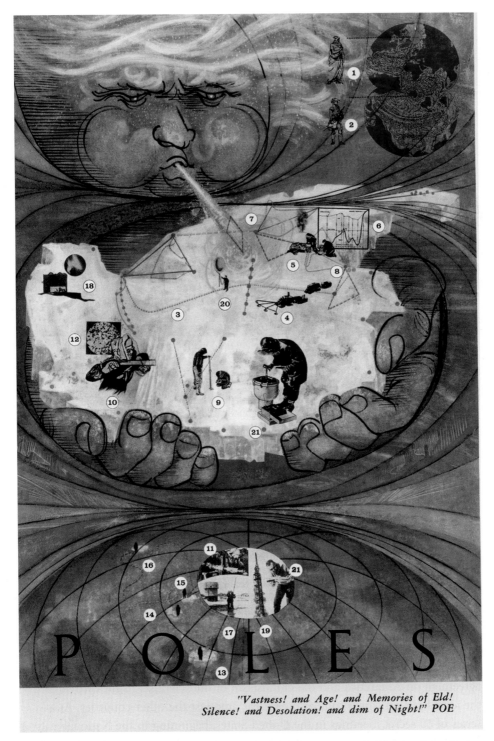

Figure 1.18 IGY poster for polar research. (Credit: National Academy of Science/IPY Office)

Figure 1.19 The first South Pole Station as it appeared upon completion of construction in January 1957. Snow drift and accumulation eventually buried the buildings, rendering it unusable by the 1970s. (Credit: US Navy, NSF)

now become a matter of international prestige so failure was not an option. Against all the odds, the stations were built, the scientists were installed and the measurements were made.

The financial investments were huge and when the scientists began to agitate for first an extension of the 'year' and then for a more permanent agreement to stay in the Antarctic they were pushing at a half open door. The United States had become concerned, even before IGY, that the USSR might wish to build military bases in Antarctica. For President Eisenhower, the IGY offered a short-term solution to the problem but a more permanent resolution was needed. During the IGY, the US government had already begun secretly to plan for an international agreement on Antarctica and the enthusiasm of the scientists to stay provided the ideal basis on which to float their ideas amongst the other countries. At a series of 60 secret meetings in the United States, beginning in the National Science Foundation Boardroom on 13 June 1958, the 12 countries hammered out the Antarctic Treaty. To emphasise joint participation, the chairmanship of the

Figure 1.20 Presidential midwinter letter from John F. Kennedy, following a tradition established in IGY. (Credit: US Navy, NSF)

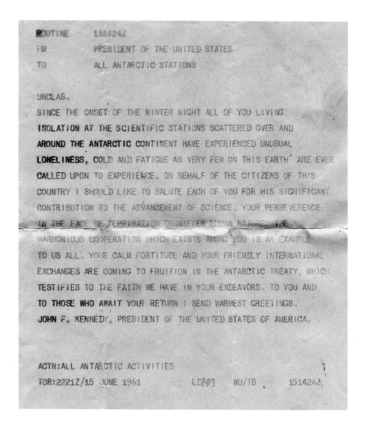

meetings rotated between countries alphabetically. The negotiations were nearly derailed several times but, with some inspired drafting and solutions to the two major political difficulties on sovereignty and non-militarisation, the Antarctic Treaty was finally signed in Washington DC on 1 December 1959. The political problems of the competing land claims were solved by setting them aside, whilst the mutual suspicions of the super powers were assuaged by the inclusion of an international inspection procedure.

International science

Whilst the physicists had always seen the Antarctic as simply a part of the world as a whole, this was not the case for many other science disciplines. Although the main thrust of IGY in Antarctica was on physical measurements a wide range of other science was undertaken. Countries such as the UK had already been researching the geology of the Antarctic Peninsula for many years, whilst others had begun biological studies on the seals and penguins, collected the lichens and mosses and measured glacier movement. As IGY faded into history, the 12 Antarctic Treaty countries began to develop their science programmes, slowly

Figure 1.21 International Polar Year logo.

diversifying into new fields as the infrastructure and transport were improved. The Scientific Committee on Antarctic Research (SCAR) provided a forum for discussing international collaboration and soon plans were hatched for ice drilling, ice thickness measurements, compilation of all the cartographic information, international symposia on Antarctic geology and biology and so on. As the momentum grew new challenges were recognised. The possibility of large-scale commercial harvesting of Antarctic krill highlighted how little was known about this key species. SCAR organised a major international programme – Biological Investigation of Marine Antarctic System and Stocks – with coordinated cruises, standardised methods and data sharing, which provided the basis for future management of the biological resources of the Southern Ocean. SCAR also provided a forum for the logistics people to plan together and exchange ideas, a forum that eventually developed into a separate organisation, the Council of Managers of National Antarctic Programs.

Working closely with the Antarctic Treaty, SCAR provided a continual flow of papers on science, conservation and environmental management, stimulating the development of most of the provisions now part of the Treaty law. As Antarctic science became more visible internationally, more countries began to show an interest in participating. Slowly, at first, new countries joined the original 12 until numbers swelled to 50 in the Antarctic Treaty (of whom 28 are currently active in the Antarctic) and 31 full members of SCAR, with a further six associate members. The Treaty Parties in total now represent around 65% of the global population, a clear majority by population even if they account for only a quarter of the nation states.

Antarctic science in the twenty-first century is marked by the same international collaboration that began 50 years ago. Indeed, to mark this and to encourage other countries to take part in polar research a new International Polar Year (IPY) was declared from March 2007 to March 2009. Over 40 000 scientists took part in projects in the Arctic and the Antarctic. This time an even wider range of disciplines were involved, ensuring not only that those who live in the polar regions had their interests taken into account but also that polar research could continue to make an increasing contribution to our understanding of how the Earth itself functions.

Imagining Antarctica

The general public views of Antarctica are either of a continent for science or of a snowy waste of little use to anyone, but threatening the world if it melts. Slowly but surely, a wider range of perceptions are growing that see the continent as a part of our culture and arts, stimulating an increasing interest from humanities scholars, fiction writers and visual artists.

Although the Antarctic has no native people, indeed no permanent population, the very idea of this polar place has excited the imagination for centuries. The early speculation on a yet-to-be found southern land where Utopia awaited the discoverer had certainly begun by 1605 whilst the growth of the 'hollow earth' theory in the eighteenth century proposed the Pole as a key point of entry. Edgar Allan Poe was a hollow-earth enthusiast and it is his *Narrative of Arthur Gordon Pym of Nantucket* in 1838 that is the most famous of these books and perhaps the first important Antarctic fiction. For many readers Poe was writing a form of science fiction, a genre first really developed for the Antarctic in Jules Verne's *An Antarctic Mystery*, whose storyline harks back to the death of Arthur Pym in the earlier novel. In his later and more famous novel *Twenty Thousand Leagues under the Sea* Captain Nemo's submarine, the *Nautilus*, visits Antarctica and after sailing beneath the ice surfaces to discover an open polar sea. There have been several other science fiction writers since but the most outstanding is Kim Stanley Robinson whose novel *Antarctica* looks at McMurdo Station in a future world as global change alters all the environmental parameters.

Antarctic poetry is largely a more recent phenomenon, although many will recognise the early contribution of Samuel Taylor Coleridge in his *Rime of the Ancient Mariner*:

And now there came both mist and snow,
And it grew wondrous cold:
And ice, mast-high, came floating by,
As green as emerald.
And through the drifts the snowy clifts
Did send a dismal sheen:
Nor shapes of men nor beasts we ken –
The ice was all between.

Both Scott and Shackleton made a point of taking poetry to the Antarctic with them – Scott's favourite was Tennyson whilst Shackleton's was Browning. There is poetry and verse of various sorts in some of the explorers' diaries and in their mid-winter magazine *Aurora Australis*. Whilst much of it is uninspiring there are occasional gems. One by Frank Debenham, a geologist with Scott, was written much later (1956) but carries real conviction and mixes religion with the human emotions.

The Quiet Land

Men are not old here
Only the rocks are old, and the sheathing ice:
Only the restless sea, chafing the frozen land.
Ever moving, matched by the ceaselessly-circling sun.

Wild birds go wandering over the face of the snow;
Bright, swift, harsh-crying, strange and heedless.
Transient in time over the mountains,
As we are transient, strangers in an old land.

Man is not old here
Creeping upon the white, brilliant brow of the world.
Less than the birds, impeded and muffled by the snow.
Unheeded by the sun, rejected by the sea.
And stunned and stunted by the silence.

Lighten our darkness, oh Lord;
And lettest thou thy servants depart in peace.
For peace is here, here in the quiet land.

And, above all, the dream is here.
The dream of this that is above all else.
Braveness and light and space, and the everlasting morning.
For this time there will be no awakening, and no journey back.
Serenity is made whole and lucid;
This time the dream will never end.

The corner is turned: we can see over the brow.
We have sought and found, and it is the land that has yielded.
The mountains gleam golden as ever, but are suddenly near,
And the peace is too deep for words. The joy
Rings out in a great echoing cry from summit to summit,
Too great for declamation: the door is open
And the walls are finally down.
To sleep here is to wake, and the resurrection
Lies in the passiveness of being one with the land.

Into the quiet land, dear Lord, we are delivered.
For here is peace, here in the quiet land.

FRANK DEBENHAM

Erebus Voices – The Mountain

I am here beside my brother, Terror.
I am the place of human error.

I am beauty and cloud, and I am sorrow;
I am tears which you will weep tomorrow.

I am the sky and the exhausting gale.
I am the place of ice. I am the debris trail.

And I am still a hand, a fingertip, a ring.
I am what there is no forgetting.

I am the one with truly broken heart.
I watched them fall, and freeze, and break apart.

BILL MANHIRE

More modern poets have also used Antarctic themes, such as the Chilean poet Pablo Neruda whose two poems – Antarctic Stones and Antarctic – were truly written from the imagination as he never visited the continent. Others, such as the New Zealand poet Bill Manhire, have spent time there and used poetry to remember both historical events from a century ago and more recent events, like the DC-10 crash on Mt Erebus.

The early explorers needed to record their discoveries, both geographical and scientific. Carrying artists and draughtsmen on board was normal and it is the artists on board the *Endeavour*, Captain James Cook's ship, that provide our earliest images of the Antarctic. Yet many expeditions did not have professional artists and it is the paintings of amateurs, many self-taught and painting simply for their own pleasure, that present very personal views of the continent. Some, of course, painted the Antarctic of their imagination, especially in attempting to visualise the heroic exploits of Scott and Shackleton, but by the Heroic Age photography had developed with its startling new images, firstly still and then later moving. For us in the twenty-first century the black and white or sepia images of a century ago have an immediate historical feel but even then there were attempts to bring colour to the landscape by hand-tinting monochrome photos. The arrival of colour photography changed yet again the visualisation of the continent, abolishing the widely held belief that there was no colour in the landscape.

Figure 1.22 Wandering albatross sketch. (Credit: Bruce Pearson)

It was the National Science Foundation that first formally recognised that there was a place for an alternative view of the continent to the scientific. Since inaugurating an Artists and Writers Program in the 1970s they have provided over 60 photographers, musicians and composers, writers and poets, painters and film makers with the opportunity to present the Antarctic through their eyes. Such has been the success of the programme that it was copied first by Australia, then by New Zealand in 1996 and most recently by Britain in 2000.

As these programmes have matured they have encompassed an increasing range of interests including jewellery and costume design, sculptors, playwrights, ceramicists and video installation artists. This many-facetted view of the continent has attracted those for whom science is a closed book, whose instinctive response to a place is emotional, and whose sensibilities lie in sounds and sights and words rather than the formalistic structures of science.

For some the elements of the landscape have provided insight into the primeval changes as land and ice interact whilst others have looked through the

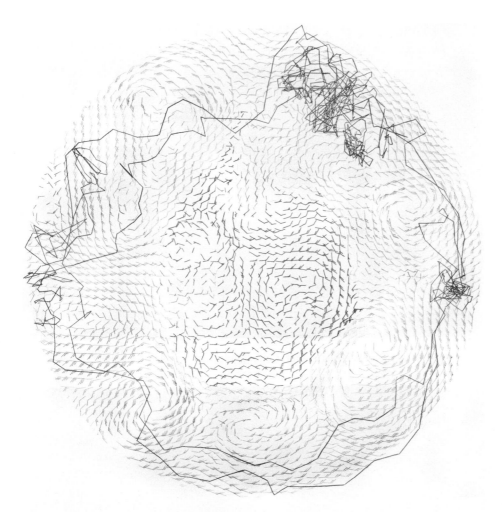

Figure 1.23 Albatross I. (Credit: Chris Drury)

eyes of scientists at hidden elements – the land beneath the ice, or the diatoms in the sea – to expose the unrecognised beauty that lies out of sight.

For others it has been the heritage from a century before, those Heroic Age huts and artefacts that remind us of brave men and an implacable land. Edward Wilson, the doctor on both of R. F. Scott's expeditions, painted delicate and evocative watercolours reminiscent of the Victorians and these, together with Herbert Ponting's remarkable photographs, have achieved an almost iconic status in the public mind. The huts still provide a focus for some modern artists and writers.

Figure 1.24 Painting of Shackleton's hut, Cape Royds. (Credit: Allan Campbell)

Figure 1.25 Flying to SkyBlu. (Credit: Philip Hughes)

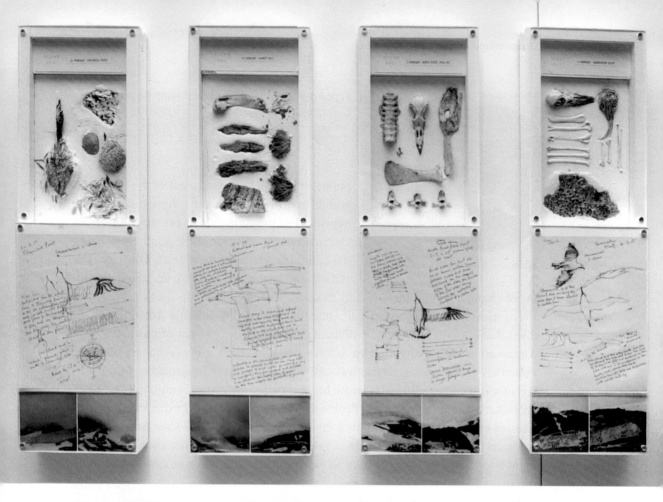

Figure 1.26 Southern Forensics – an art piece made from found objects, sketches and narrative on Signy Island. (Credit: John Kelly)

These days the Antarctic artists come with many different skills and backgrounds. Some, such as Philip Hughes, were originally engineers or scientists and from them can come an analytical clarity of vision that dissects and abstracts the landscape into its basic elements. Others with interests in geology and geography, such as John Kelly, see the landscape through its detritus, the ice-smoothed stones, the discarded penguin feathers, the broken petrel egg. Lucia de Leiris, an American natural history artist, has much in common with Edward Wilson in her portrayal of birds and seals, whilst some of the scientists themselves have remarkable artistic talents such as Richard Laws, whose zoological expertise lends accuracy to his paintings of elephant seal behaviour.

The new art has no defining characteristics as the practitioners come from many countries and backgrounds. The common theme is the experience of being in the Antarctic and finding new ways to provide that emotional high to a wider

Figure 1.27 South Pole marker. (Credit: Glenn Grant, NSF)

public. It is both the strangeness of the place and its international basis that have provided new ideas to these artists. For example, Xavier Cortada's flag installation at the South Pole marks the natural annual movement of the Pole since the station was established in 1956, yet also flags up major changes around the world in those 51 years.

For many people the archetypal Antarctic music is the Sinfonia Antarctica (Symphony No. 7) by Vaughan Williams, developed in 1952 out of music he composed to accompany the film *Scott of the Antarctic*. Yet the composer never visited the polar regions, so this was entirely a work of the imagination. More recently Peter Maxwell Davies wrote his *Antarctic Symphony* (Symphony No. 8) only after visiting the Antarctic Peninsula and gaining personal experience of this unusual environment. Other composers in New Zealand (Chris Cree Brown), Australia (Nigel Westlake), Britain (Craig Vear, Muriel Johnstone) and Canada (Ian Tamblyn) as well as Vangelis have all provided new and different approaches to Antarctic music.

Now there are other ways of experiencing the Antarctic. Tourism, which began around 50 years ago, now provides an opportunity for over 35 000 people a year to visit the Antarctic and see for themselves, however briefly, the majesty of the white continent. Managing this comparatively large number of people who all come in summer and whose visits are largely restricted to fewer than 60 sites is proving a continuing political and environmental problem for the Treaty Parties. And the familiarity of penguins in all their various forms, from cartoons to wallpaper, provides an international and accessible icon for the continent.

The release of satellite information on Google Earth and webcams on Antarctic stations has also changed the way everyone can interact with these frozen wastes. Indeed, this visualisation of Antarctica through the internet, TV and newspapers, as well as through feature films, has made the continent suddenly much more familiar to the public than ever before. It directly links with global change and therefore political concerns of the highest importance have ensured increasing coverage of its scientific output by all the media outlets, and it has even appeared in the educational systems of several countries as a part of the standard curriculum.

Figure 1.28 Tourists ski by the ceremonial pole at Amundsen–Scott South Pole Station as preparations begin to honour the centennial of the first people to arrive at the geographic South Pole on 14 December 1912. (Credit: Brite Niezek, NSF)

Antarctica is no longer outside of our experience or beyond our comprehension. It is increasingly seen as important to our future, a key element of how the Earth functions and a space for portraying human emotions through the many forms of art.

List of principal Antarctic expeditions

1772–75	J. Cook	British naval expedition	*Resolution*
1822–24	J. Weddell	British sealing voyage	*Jane*
1826–29	J.-S. Dumont d'Urville	French naval expedition	*Astrolabe*
1828–31	H. Forster	British naval expedition	*Chanticleer*
1838–42	C. Wilkes	US Exploring Expedition	*Vincennes*
1839–43	J. C. Ross	British naval expedition	*Erebus* and *Terror*
1872–76	C. W. Thomson	British naval voyage	*Challenger*
1892–93	A. Fairweather	British whaling expedition	*Balaena*
1893–94	C. A. Larsen	Norwegian sealing and whaling	*Jason*
1893–95	H. J. Bull	Norwegian sealing and whaling	*Antarctic*
1897–99	A. de Gerlache	Belgian Antarctic Expedition	*Belgica*
1898–1900	C. Borchgrevink	British Antarctic Expedition	*Southern Cross*
1901–04	R. F. Scott	British National Antarctic Expedition	*Discovery*
1901–03	E. von Drygalski	1st German South Polar Expedition	*Gauss*
1901–04	O. Nordenskjöld	Swedish South Polar Expedition	*Antarctic*
1902–04	W. S Bruce	Scottish National Antarctic Expedition	*Scotia*
1903–05	J. B. Charcot	1st French Antarctic Expedition	*Français*
1904		Establishment of Argentine station at Laurie Island	
1907–09	E. Shackleton	British Antarctic Expedition	*Nimrod*
1908–10	J. B. Charcot	2nd French Antarctic Expedition	*Porquois-Pas?*
1910–12	R. Amundsen	Norwegian Antarctic Expedition	*Fram*
1910–13	R. F. Scott	British Antarctic Expedition	*Terra Nova*
1911–12	N. Shirase	Japanese Antarctic Expedition	*Kainan-Maru*
1911–12	W. Filchner	2nd German South Polar Expedition	*Deutschland*
1911–14	D. Mawson	Australian Antarctic Expedition	*Aurora*
1914–16	E. Shackleton	British Imperial Trans-Antarctic Expedition	*Endurance*
1920–22	J. L. Cope	British Expedition to Graham Land	
1921–22	E. Shackleton	Shackleton–Rowett Antarctic Expedition	*Quest*
1923–24	C. A. Larsen	Norwegian whaling expedition	*Sir James Clark Ross*
1925–27	S. W. Kemp	Discovery Investigations	*Discovery*
1927–28	H. Mosby	Norwegian Antarctic Expedition	*Norvegia*
1928–30	R. E. Byrd	United States Antarctic Expedition	*City of New York*
1929–31	D. Mawson	BANZARE	*Discovery*
1933–35	R. E. Byrd	United States Antarctic Expedition	*Bear of Oakland*
1934–37	J. R. Rymill	British Graham Land Expedition	*Penola*
1938–39	A. Ritscher	German Antarctic Expedition	*Schwabenland*
1939–41	R. E. Byrd	US Antarctic Service Expedition	*North Star*

Note: A much fuller list of all Antarctic expeditions can be found in R. K. Headland (2009). *Chronological List of Antarctic Expeditions and Related Historical Events*. London: Quaritch.

2 | A keystone in a changing world

BRYAN STOREY

Ferrar had returned all right and had made a great find in the western mountains of fossil plants in some coal shale beds which means that the South Pole has at one time had an abundant vegetation. **Edward Wilson, 1902**

In contrast to today, Antarctica has not always been the cold, isolated continent sitting astride the South Pole. The mountains that protrude through the thick veneer of ice provide a past record that is very different to today, an intriguing glimpse of a continent that has preserved a remarkable record of how our Earth has changed through geological time, and how Antarctica amalgamated and combined with other continental fragments and drifted apart to form new ocean seaways. Despite Antarctica being almost completely covered in ice, with less than 0.5% of the continent exposed rock, we know a great deal about how the Antarctic continent has changed through time from the oldest rock, estimated by geologists to be about 3930 million years old, to the geological processes that are operating today. Our understanding has developed through the tireless efforts of early explorers who paved the way for modern day geologists visiting and sampling all accessible rock outcrops to unravel the mysteries of the hidden continent. Initially geologists travelled from rocky outcrop to rocky outcrop by sledge using reliable and trustworthy dog teams to be replaced in later years by mechanised motor toboggans and air support.

Geologists working in Antarctica have provided important knowledge and data in support of the theory of plate tectonics. They discovered fossil plants and animals that matched similar finds in countries that were once part of the much larger Gondwana supercontinent. Our knowledge of Antarctica has been further enhanced by developing geophysical and remote sensing techniques that provide an insight into the geology and form of the continent beneath the ice. These techniques include

Antarctica: Global Science from a Frozen Continent, ed. David W. H. Walton. Published by Cambridge University Press. © Cambridge University Press 2013.

Figure 2.1 Geologist studying a large volcanic dyke on Saunders Island. (Credit: Pete Bucktrout, BAS)

Figure 2.2 Geologist drilling samples from outcrop. (Credit: Brian Thomas, BAS)

Figure 2.3 In the early days of geological exploration of Antarctica geologists used dog teams to travel from one location to another collecting rock samples. This photo shows Lieutenant Jack Tuck (US Navy) on the annual sea ice near what is today known as McMurdo Station. (Credit: US Navy, NSF)

airborne magnetic and gravity surveys that detect the different physical properties of the rocks below, and oversnow seismic techniques that image reflective horizons within geological strata. In recent years, major international consortia have been able to drill down through the ice into the Earth's outer crustal layer to sample rocks up to 1170 m below the ground surface. A recent international ANDRILL (Antarctic Geological Drilling) project has brought together scientists from the United States, New Zealand, Germany and Italy to develop modern drilling technology. Their rock cores have provided a unique record of how Antarctic climate has changed over the last 30 to 40 million years, from close to the time when Antarctica, although still at the South Pole, was part of a greenhouse world devoid of ice. This information is critical for measuring global climate change, now seemingly accelerating.

There are many isolated rocky mountain peaks, known as nunataks, that protrude through the icy carapace, particularly near the periphery of the Antarctic continent. But it is three spectacular mountain ranges that provide the most amazing insights into the geology of the frozen continent: the Transantarctic Mountains, a high

Figure 2.4 British Antarctic Survey Twin Otter equipped to fly a geophysical survey in Antarctica to investigate the rocky landscape beneath the ice. (Credit: David Vaughan, BAS)

Figure 2.5 Drilling rig on sea ice off Cape Roberts, December 1999. (Credit: John Smellie, BAS)

mountain range that traverses the continent dividing it into an older 'East Antarctica' from a younger 'West Antarctica'; the Ellsworth Mountains, the highest mountain range that occupies what appears to be an anomalous or unusual position aligned transverse to the linear trend of the Transantarctic Mountains; and the Antarctic Peninsula, a deeply dissected, highly glaciated, sinuous mountain chain.

Figure 2.6 The Transantarctic Mountains, glaciers and crevasse fields. (Credit: Bryan Storey)

Figure 2.7 Seismic field camp on the Rutford Ice Stream with the Ellsworth Mountains beyond. (Credit: Andy M. Smith, BAS)

A hidden landscape

If we were able to see beneath the ice, a very different landscape would be
revealed. The lowlands that characterise East Antarctica are interrupted by
subglacial highlands, the Gamburtsev Mountains that reach up to only 600 m
beneath the surface of the 4 km thick ice sheet. The highlands are about the same
size as the European Alps, about 1200 km long and about 3400 m from base to
the highest point. Not only have these mountains never been seen but we can only
speculate as to why they should occur in such an unusual position in the centre of an
old continental shield. The Gamburtsev Mountains may represent the eroded
remnants of a very old mountain range, a view that is in keeping with the known
history of East Antarctica yet in contrast with other ancient continental interiors
such as Australia or North America where there are no such central old
highland regions that have survived millions of years of weathering and erosion.

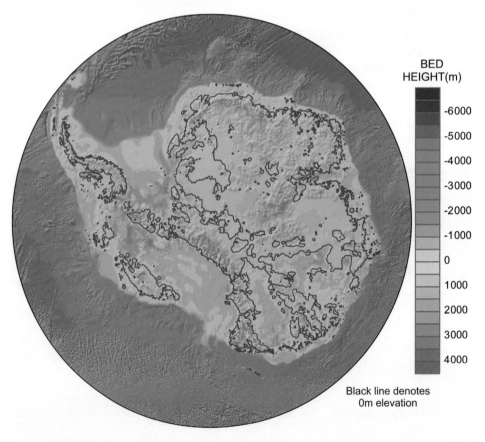

Figure 2.8 Sub-ice topography map showing the rock underlying the continental ice sheet.
(Copyright SCAR)

Alternatively, and a more exciting possibility, they may represent a volcanic uplands currently or recently active that sits above an anomalously hot part of the Earth's interior or close to a previously unrecognised boundary between two portions of the Earth's outer crust, known as tectonic plates, that have moved relative to each other. The presence of such a large hidden highland region is in marked contrast to West Antarctica where much of the landscape is below sea level with the Byrd Subglacial Basin and the sinuous Bentley Subglacial Trough plummeting down to 2000 m below sea level, that is, after we take the weight of the Antarctic ice sheet into account. If the ice sheets were to melt much of West Antarctica would be flooded by seawater.

Although the ice sheets are, for the most part, sitting on and moving slowly across a rocky substrate we now know that more than 300 lakes exist beneath the ice sheets. The largest and most well known is Lake Vostok that sits in a depression close to the subglacial Gamburtsev Mountains. It is approximately 250 km long, 50 km wide at its widest point and has an average depth of 344 m. Some of the lakes are apparently connected by sub-ice channels whereas others remain isolated. The subglacial lakes have to this day remained unexplored although scientists are keen to probe the hidden depths of the lakes and test the widely held view that life exists in what is predicted to be an unusual geochemical environment.

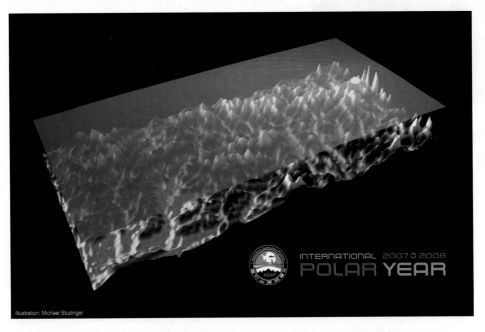

Figure 2.9 Gamburtsev Mountains. A three-dimensional image constructed from radio-echosounding data. (Credit: Michael Studinger)

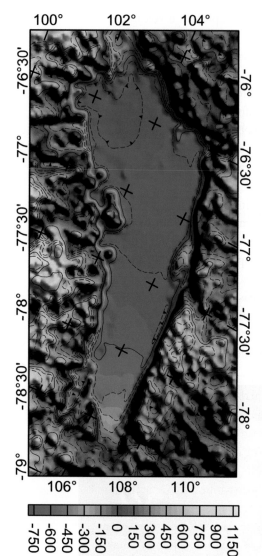

Figure 2.10 Map of Lake Vostok, the largest subglacial lake lying in a rift in the Gamburtsev Mountains. (Credit: Michael Studinger)

A sleeping giant

Unlike the other continents, Antarctica has low seismic activity and very few active volcanoes. It was originally thought that the thick ice sheets may have masked any earthquake activity but we now know it is a characteristic feature of Antarctica, with the exception of the northern portion of the Antarctic Peninsula. As far as we know, the Antarctic continent is not crossed by or close to any plate boundaries, which is where we expect to find volcanic and earthquake activity. Antarctica sits in the middle of a large plate – the Antarctic Plate –surrounded by constructive [growing], diverging plate boundaries beneath the Southern Ocean. This means that earthquakes and volcanic activity occur well away from the Antarctic continent and that the plate is growing at its edges, moving away from neighbouring plates with new basaltic rock being added to the leading edge of the plate. This contrasts with the northern tip of the Antarctic Peninsula where a series of small tectonic plates are caught up between the big Antarctic and Southern American plates. One small plate, the Drake Plate, is currently sinking very slowly (with associated seismic activity) and being destroyed beneath the South Shetland Islands, a process known as subduction. The Drake Plate is the last remaining part of what was once a much larger plate that was subducting beneath the western margin of the whole Antarctic Peninsula. With the exception of steaming fumaroles on Bridgeman Island, there is no volcanic activity on the South Shetland Islands associated with the subducting Drake Plate. However, a marine basin forming the current seaway of Bransfield Strait, opened a few million years ago between the South Shetland Islands and the Antarctic Peninsula, with Deception Island – one of the two currently most active volcanoes in Antarctica – situated on the spreading ridge, in some ways analogous to Iceland sitting on the Mid-Atlantic

Figure 2.11 Map showing the Antarctic tectonic plate and its interactions with other southern hemisphere plates including the Scotia Sea region, based on satellite gravity measurements. (Credit: Trond Torsvik)

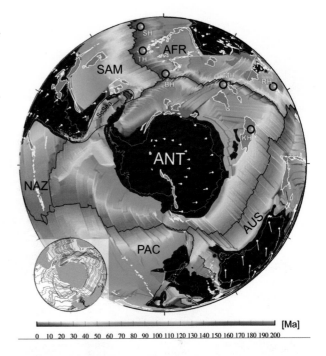

spreading ridge. Deception Island had significant eruptions in 1967, 1969 and 1970. The island is around 14 km in diameter, and breached at the southeast end allowing seawater to fill the central caldera. Three new craters and an island were formed in the 1967 eruption. Volcanic ejecta were erupted from a fissure in the ice on the caldera wall in 1969 destroying a Chilean station and severely damaging a nearby British station. A chain of new craters was formed in 1970. Hot springs near the shoreline of the caldera provide a perfect location for swimming and bathing and have become a popular destination for tourists on Antarctic cruises.

The most active volcano in Antarctica is Mt Erebus, the southernmost active volcano in the world and the largest (3795 m high) of four volcanic cones that form Ross Island. It contains a persistent lake of molten magma that produces an unusual volcanic rock popularly called kenyte (technically known as anorthoclase phonolite, a unique type of feldspar mineral). The only other place in the world where this volcanic rock is found is in Kenya. Although the last major historically known eruption occurred in 1984–85, small relatively low-level sporadic volcanic eruptions to altitudes of tens to hundreds of metres are common from the lava lake today. Interestingly, Mt Erebus does not sit on nor is it related to an active plate boundary, but it is the remaining active volcano in a much larger volcanic province, the McMurdo volcanic province, that was caused by, or in some way related to, rifting of the continent that resulted in large parts of West Antarctic being below sea level and uplift of the Transantarctic Mountains on the flanks of the rift system.

Figure 2.12 Deception Island, South Shetland Islands, showing the 10 km diameter flooded caldera. The caldera collapsed approximately 10 000 years ago following a major eruption. Several smaller eruptions occurred between 1967 and 1970. (Credit: John Smellie, BAS)

Figure 2.13 (a). A view of the old, inactive crater (foreground) and main crater of Mt Erebus, the southernmost active volcano in the world on Ross Island. (b) Closeup of the Erbus main crater, which contains an active lava pool. (Credits: Antarctica New Zealand)

In 1992 the Carnegie Mellon University designed and developed a robot 'Dante' to enter and study the unique convecting magma lake inside the Mt Erebus inner crater. Dante was designed as a tethered walking robot capable of climbing steep slopes and self-sustained operation in the harsh Antarctic climate. Unfortunately, a break in a fibre optic cable linking the 8-foot (2.4 m) high spidery robot to the control hut halted the mission as Dante made its way down the steep and treacherous crater walls. The expedition demonstrated the advancing state-of-art in mobile robotics and the future potential of robotic explorers in remote locations such as Antarctica.

As is always the case in Antarctica, many exciting discoveries remain hidden beneath the ice. Scientists have recently, and for the first time, discovered exciting evidence for a volcano completely covered by the thick Antarctic ice sheet. They, using ice-penetrating radar, have estimated that the volcano last erupted about 2300 years ago, yet remains active today. This was probably the biggest eruption in Antarctica during the last 10 000 years. It blew a substantial hole in the ice sheet close to the Pine Island Glacier on the West Antarctic icesheet and generated a plume of ash and gas that rose 12 km into the air, covering an area of 20 000 km^2. Although ice has buried the unnamed volcano, molten rock is still erupting below the ice. It remains to be seen how many more such volcanoes exist beneath the ice and whether our concept of a sleeping giant is correct. Like Mt Erebus, the buried volcano does not relate to a plate boundary but may be the last vestige of a previously active rifted landscape that had many more active volcanoes than today.

Birth of Gondwana

In contrast to the present isolated position of Antarctica astride the South Pole, the continent has amalgamated and combined with other continental fragments and rifted apart to form new ocean seaways throughout geological time. For much of the Phanerozoic (from 540 to 180 million years ago) Antarctica was the keystone of the Gondwana supercontinent. It was joined centrally to South America, India, Africa, Australia and New Zealand. More amazingly, prior to Gondwana, many scientists have speculated that East Antarctica may have been joined to the western side of America in a continent called Rodinia. The hypothesis has become known as the SWEAT hypothesis (South West US–East Antarctica Connection). As North America left East Antarctica scattered continental blocks came together again to form Gondwana about 550 million years ago at a time when there was rapid evolution of multicellular plants and animals. Where adjacent continental blocks collided, Himalayan-sized mountain chains traversed the continent – the deeply eroded remnants of these sutures can be seen today in the

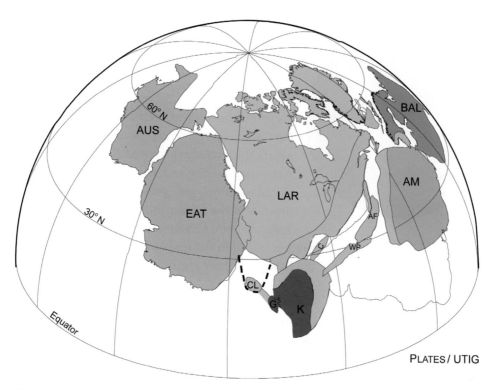

PLATES / UTIG

Figure 2.14 A continental reconstruction for a supercontinent named Rodinia showing how East Antarctica (EAT) was most likely joined to North America (LAR) 750 million years ago. (Credit: Ian Dalziel)

scattered nunataks that protrude through the East Antarctic ice sheet. Subduction of proto-Pacific ocean floor was also taking place along the length of the Transantarctic Mountains at this time, forming many of the rocks that we see today spectacularly exposed in the uplifted mountain range.

Antarctica: a keystone of ancient Gondwana

For the 300 million years of its history, Gondwana did not remain in a fixed position, but wandered from a more equatorial position southwards to a more polar position as the supercontinent started to disintegrate. It migrated through different climatic regions and was colonised by an evolving flora and fauna.

For the first part of its journey, picture a continent traversed by large meandering river systems and partly covered in lakes and shallow seas inhabited by armour-plated fish, now preserved as fossils in sedimentary rocks along the length of the Transantarctic Mountains. Inevitably, climatic conditions changed dramatically and just as the present continent is covered in ice, so too was the

Carboniferous Ice World
289 Ma

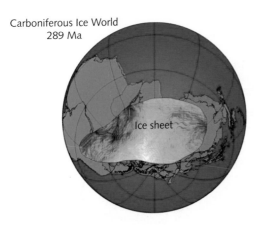

Ice sheet

Figure 2.15 The Carboniferous Ice World showing how a large ice sheet covered much of the Gondwana supercontinent about 300 million years ago. (Credit: Bryan Storey)

ancient Gondwana supercontinent during the Carboniferous period about 300 million years ago. A thick ice sheet covered much if not all of Antarctica, waxing and waning and depositing glacial moraines, now preserved as thick glacial deposits in the Ellsworth Mountains and in the Transantarctic Mountains. As the ice moved over the underlying land surface, boulders locked in the ice cut furrows and grooves, providing a further record of the passage of these great ice sheets long after they disappeared.

After the Carboniferous ice world, climate warmed and there was rapid evolution of a rich flora. Extensive cool-temperate swamps with thriving plant communities formed coal deposits, which can now be seen as thick coal seams stretching along the length of the Transantarctic Mountains. In fact, over 100 years ago, Edward Wilson, as he made his way up the Beardmore Glacier as a member of Scott's Polar party, realised from these conspicuous coal horizons and from finding fossil leaves that Antarctica had been very different from today and not always the cold barren icy landscape that he was experiencing. Some of the most beautiful and distinctive fossils entombed in these Gondwana coal measures are the leaves of the ancient deciduous tree, *Glossopteris*. Early Ginkgos and conifers also appeared in the fossil record in Antarctica and the once neighbouring continents of Gondwana (e.g. Australia, India, etc.). They colonised the drier hillsides and were not restricted to swamps. Tree ferns similar to those alive today were abundant. The climate warmed up gradually after the ice age. Nevertheless it was a cold climate for much of this period. The Permian *Glossopteris* flora is one of the most characteristic fossil assemblages of the Gondwana supercontinent. The discovery of these fossils on the now widely distributed Gondwana continents of South America, Africa, Australia and India provides important supporting evidence for the theory of continental drift and plate tectonics. Unfortunately, you will not find any descendants of the glossopterids alive today. They were wiped out by the end-Permian extinction event, the Earth's most severe extinction event, with up to 96% of all marine species and 70% of terrestrial vertebrate species becoming extinct. There are several proposed mechanisms for the extinctions; an earlier peak was likely due to gradual environmental change (including sea-level change, anoxia and increased aridity), while a later peak was probably due to a catastrophic event

Figure 2.16 A thick coal seam from high in the Transantarctic Mountains formed when Antarctica was linked to Gondwana and located in a different mid-latitude climatic belt to today. (Credit: Bryan Storey)

Figure 2.17 (a) Fossil leaf from the extinct *Glossopteris* tree. (Credit: David Cantrill) (b) Fossil fern from the Lower Cretaceous, Alexander Island. (Credit: Chris Gilbert, BAS) (c) Cretaceous fossil bivalve *Aucellina* from Alexander Island. (Credits: Chris Gilbert, BAS)

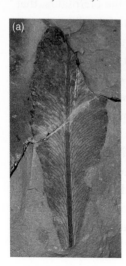

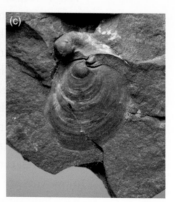

such as large or multiple meteorite impact events, increased volcanism or sudden release of methane hydrates from the sea floor.

A warm and wet interval followed the end Permian extinction with the sudden appearance of a new flora that is currently represented in the present day flora of the once neighbouring continents of South America, Australia and New Zealand. The new flora was characterised by the first forked-frond seed ferns, podocarps, conifers, ferns, ginkgos, horsetails and cycadophytes. The climate became increasingly hot and dry and many of the plants show drought-resistant adaptations. *Dichroidium* fern is the characteristic plant of this period. Plants are not the only fossils that tie the disparate parts of Gondwana together. So too do numerous vertebrate fossils, including a diverse assemblage of therapsid reptiles *Lystrosaurus*, *Myosaurus*, *Thrinaxodon* amongst others. *Lystrosaurus* is a typical vertebrate that occurs in many other Gondwana continents in addition to Antarctica. This animal was a dicynodont, a mammal-like reptile, with a distinctive skull that possessed two impressive tusks. It roamed the fluvial flood plains feeding on the lush vegetation. *Myosaurus* is a very small rare dicynodont known only from South Africa and Antarctica, and *Thrinaxodon* is a weasel-sized carnivorous cynodont that also occurs in South Africa. Occurring with the therapsids are a primitive group, the prolacertids, related to modern lizards, and the temnospondyls, a fairly diverse group of extinct semi-aquatic amphibians ranging in length from less than a metre to the size of a large crocodile with skull lengths of nearly 1 m.

The cynodonts were replaced by the carnivorous thecodonts, which later gave rise to the dinosaurs. One dinosaur, *Cryoloophosaurus ellioti*, is a 7 m long carnivorous theropod dinosaur unique to Antarctica. The name means 'frozen crested reptile' in reference to its polar occurrence and the unusual bony display crest on top of the skull. Large long-necked sauropod herbivores such as *Apatosaurus* and *Brachiosaurus* were also present along with the flying reptiles, the pterosaurs, that co-existed with the dinosaurs during the Mesozoic. The success of the dinosaurs was quick and striking, and seems to have been associated with equally striking climate and floral changes. The *Dicroidium* flora gave way to a conifer-dominated plant assemblage, which was to flourish for much of the period of the dinosaurs. Only near the end of the dinosaurs' long and successful reign did the flowering plants begin their spectacular rise to prominence, culminating in their dominance in the last 70 million years, the Cenozoic.

On the Antarctic Peninsula side of the continent, younger, 70 million year old vertebrates, carnivorous plesiosaurs and mosasaurs have been found. Plesiosaurs and mosasaurs are extinct marine reptiles that co-existed with the dinosaurs, the former have long serpentine necks, whereas the latter are closely related to modern lizards. In addition to these marine animals, a few fragmentary dinosaur fossils have been found including a nodosaurid (an armoured herbivore related to the ankylosaurs), a small herbivorous hypsilophodontid, and hadrosaurus, the larger herbivorous duck-billed dinosaur. The diverse and interesting fossil history of Antarctica would not be complete without mentioning the long invertebrate fossil record known from the Lower Cambrian (about 500 million years ago) onwards. For the most part they are marine in origin with rare occurrences of freshwater and terrestrial species, archaeocyathids, trilobites, molluscs, brachiopods, crinoids, echinoderms, corals, serpulids, crustaceans and bryozoans.

The fragmentary fossil record from Antarctica indicates that a surprisingly wide variety of organisms lived on or around Antarctica. The southern continent has been colonised by a succession of plants and animals that shows signs of being as complex as those on any other continent. Vast habitat areas for occupation by both terrestrial and marine organisms, coupled with equitable climates for very long periods of time, have contributed towards this proliferation of life. Antarctica seems to have been a particularly important dispersal route and undoubtedly served as a major corridor for floral and faunal interchange between the high and low southern latitudes. A number of plant and animal groups seem to have both originated in, and then radiated from the high southern latitudes. Even after the marked climate cooling and loss of habitats through the Cenozoic (last 65 million years) there still appears to have been adaptive radiations especially in the marine realm. It is possible that temperate, cool temperate and even cold temperate regions of the world have been very effective in the process of species diversification. Continued investigations of the fossil record may provide valuable insights into the role of the polar regions in determining the contribution of high latitude regions to the global species pool.

A land of fire

The relative peace and tranquillity of Gondwana was disrupted approximately 183 million years ago by an extensive but relatively short-lived period of volcanic activity that heralded the break-up and disintegration of the supercontinent. The intensive volcanism, which lasted for less than a million years, stretched from what was to become Southern Africa into Antarctica, along the length of the Transantarctic Mountains, and on into Tasmania and New Zealand, a distance over 3000 km. Although much of the volcanic debris that was ejected from numerous volcanoes has been eroded, we are left with an impressive network of subvolcanic conduits, mostly horizontal (sills) but some vertical (dykes) layers that record the passage of hot magma from its source within the Earth's mantle. Today, if you stand in the Transantarctic Mountains, you cannot but be impressed by the lateral continuity of these sills that stretch for literally hundreds of kilometres along the steep escarpment. In the Theron Mountains, more than 20 such sills form over

Figure 2.18 Basaltic sills in the Theron Mountains illustrating the large volume of subvolcanic magma emplaced along the length of the Transantarctic Mountains and in neighbouring Gondwana continents just prior to the disintegration of Gondwana. (Credit: Bryan Storey)

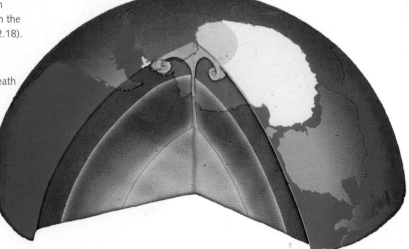

Figure 2.19 A Ferrar basaltic sill in the Dry Valleys over 2500 km away from similar basaltic sills in the Theron Mountains (see Figure 2.18). (Credit: Bryan Storey)

Figure 2.20 Mantle plume beneath Gondwana. The mantle plume was most likely responsible for the extensive basaltic volcanism and may have contributed to the break up of Gondwana. (Credit: Bryan Storey)

Figure 2.21 A reconstruction
of Gondwana 180 million years
ago (Ma) and its subsequent
fragmentation. The red dots show the
location of possible mantle hot spots
that may have contributed to the
break up of Gondwana. (Credit: Bryan
Storey)

(a)

Gondwana 180 Ma

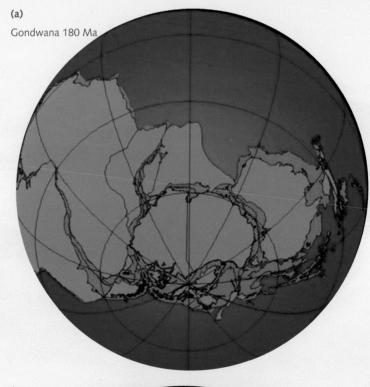

(b)

South Atlantic
opening
130 Ma

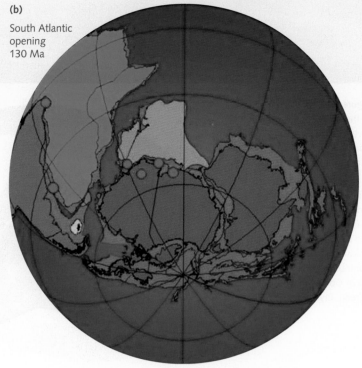

(c)

Australia and
New Zealand rifting
100 Ma

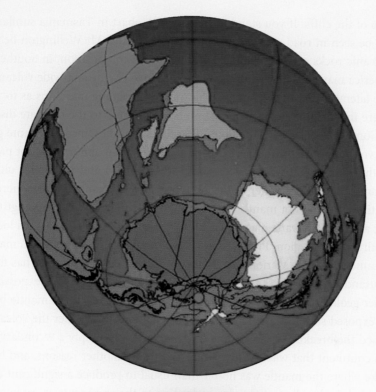

(d)

Isolation of Antarctica
35 Ma

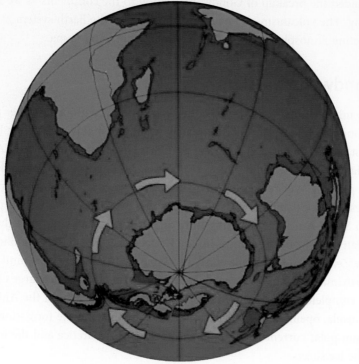

50% of the cliffs. If you stand in the city of Hobart in Tasmania similar sills can be seen in road cuttings or in the rocky cliffs of Mt Wellington behind. Volcanic rocks of a similar age, but different in composition, in Southern South America may even be part of the same extensive Gondwana-wide volcanic province. The lateral extent of the province has raised intriguing questions as to why and where this magma was generated and how it related to the ultimate disintegration of Gondwana. There is little doubt that the Karoo Province, the name given to the volcanic rocks in southern Africa, came from an abnormally hot part of the Earth's mantle that was located exactly where Africa and Antarctica ultimately separated. Geologists refer to this as a 'hot spot' or mantle plume where partial melting of rising hot mantle forms the volcanic magma. Some geologists suggest that all the Gondwana magma came from a couple of those hotspots located in this South Atlantic region. Present day representation of these hot spots may be the volcanic islands of Marion and Bouvet. If this is the case, magma has flowed for thousands of kilometres through the dykes and sills that we see exposed today. Other geologists prefer a local source for these magmas in the mantle beneath the exposed sills and dykes. We are also unsure as to whether the volcanic activity caused the breakup of Gondwana or whether it was simply a secondary response to a continent that was already rifting apart due to other reasons, and by chance broke where the mantle was hotter than usual to produce a significant event that heralded the breakup of Gondwana leading to the continents as we know them today. The volcanism also had a major effect on the Earth system at that time affecting atmospheric composition, climate, flora and fauna.

Gondwana breakup

The initial rifting stage started either synchronous with or just after the outburst of volcanic activity ultimately leading to a seaway forming between West (South America and Africa) and East Gondwana (Antarctica, Australia, India and New Zealand) and to sea floor spreading in the Somali, Mozambique and possibly Weddell Sea basins. The second stage occurred in Early Cretaceous times (circa 130 million years ago) when this two-plate system was replaced by three with South America separating from an African-Indian plate, and the African-Indian plate from Antarctica. In late Cretaceous times (90–100 million years ago), New Zealand and Australia started to separate from Antarctica until finally at approximately 32 million years ago, the breakup of that once large continent was complete when the tip of South America separated from the Antarctica Peninsula, opening up the Drake Passage and allowing the formation of the circum polar current resulting in the cooling of Antarctica and the formation of the ice sheets.

Figure 2.22 Basaltic dykes and sills in Marie Byrd Land related to volcanic activity active when New Zealand rifted from Antarctica (Credit: Bryan Storey)

Intriguingly, most of the Gondwana breakup stages described above were associated with intense periods of volcanic activity. The opening of the South Atlantic resulted in the separation of South America from Africa. Two present day volcanic islands, Tristan da Cunha and St Helena, are connected via submarine ridges to large volcanic provinces, the Parana basaltic province in South America and the Etendeka province in Africa, respectively. Although we know the volcanic provinces are in some way related to continental breakup, the cause and effect are still matters of debate and ongoing research.

A jigsaw puzzle: rotating microcontinents

Whatever processes controlled or influenced the breakup of Gondwana, for the most part the supercontinent separated into the major continents as we know them today. However, some small microplates were also formed, particularly in the South Atlantic region, and these help to explain some of the geographic features of Antarctica, particularly the Ellsworth Mountains. An Ellsworth microplate, formed at the time of breakup, rotated clockwise more than 90° and migrated from an original position between South Africa and Antarctica to its present position at the head of the Weddell Sea. This explains the anomalous direction of the Ellsworth Mountains, relative to the Transantarctic Mountains. It also explains the anomalous position of a small group of nunataks, Haag Nunataks, discovered in the 1980s to be old metamorphic rocks from East Antarctica that formed approximately 1200 million years ago. They too represent an exotic block that now forms part of West Antarctica. The Falkland Islands, like the Ellsworth Mountains, rotated 180° from a similar position between Africa and Antarctica to join South America on the opposite side of the Atlantic. We cannot be sure exactly how and why these microplates and rotations occurred or why Gondwana should have broken up in this way, but many suspect that it is in some way linked to the formation of the volcanic province at the start of rifting. The trajectory, displacement history and rotation mechanisms of the microplate component of the Gondwana jigsaw remain a mystery.

Interestingly, displaced microplates are also a significant component of the final separation of South America from the Antarctic Peninsula and the opening of Drake

Figure 2.23 Geophysicist reading a gravimeter at Haag Nunataks, an isolated exposure of Precambrian rock. (Credit: BAS)

Passage in the Scotia Sea region. The South Georgia block has moved from a position much closer to southern South America to its present position; the South Orkney block separated from the northern tip of the peninsula to form Powell Basin, and the South Shetland Islands have separated from the peninsula by sea floor spreading in Bransfield Strait. Although the driving mechanisms may not always be the same, the history of rotated microplates has been repeated in the South Atlantic region and raises intriguing questions about breakup processes. The opening of Drake Passage has also led to the formation of the South Sandwich Islands, an isolated chain of 11 volcanic islands in the South Atlantic sector of the Southern Ocean, and one of the Earth's classic and best preserved arc trench subduction systems formed entirely of volcanic rocks on oceanic crust devoid of any continental influence. The volcanic arc is tectonically simple being located on the small oceanic Sandwich Plate, which is overriding the southernmost part of the South American Plate at the South Sandwich Trench. The 11 islands are small, typically 2–12 km across and entirely volcanic in origin. They typically rise 500–1000 m above sea level and are subaerial summits of edifices that rise some 3 km above the surrounding seafloor. The arc is currently volcanically active with intense fumarolic activity on several of the islands and there have been at least six historic eruptions. Summit calderas are present on several of the islands with small persistent lava lakes

Figure 2.24 Volcanic crater at the summit of Mt Sourabaya of Bristol Island, South Sandwich Islands. Snow-covered peak in the background is an older inactive volcanic centre (Mt Darnley). (Credit: Pete Bucktrout, BAS)

frequently present. Montagu Island, with no previous record of historical eruptions, began a long-lived eruption in late 2001 and was still erupting in June 2005. There are some adjacent volcanically active submarine sea mounts with black smokers erupting beneath the sea surface.

Ring of fire

Whilst Gondwana was undergoing a phased breakup, Pacific Ocean floor was continually being subducted during the Mesozoic period beneath the Pacific margin of Gondwana, which was to become New Zealand, the Antarctic Peninsula and west coast of South America. The subduction process, which may incidentally have played its part in the breakup of Gondwana, resulted in sedimentary material being scraped off the down-going slab and added on (accreted) to the western margin of

Figure 2.25 Rocks that have been folded and scraped off the down-going slab and accreted to the western edge of the Antarctic Peninsula during the subduction process. (Credit: Bryan Storey)

Gondwana, and an active volcanic mountain chain. The eroded remnants of these volcanoes can still be seen in the deeply dissected mountains of the spine of the Antarctic Peninsula. With the exception of the northern tip of the Antarctic Peninsula, the subduction process has stopped along the Antarctic Peninsula because the spreading ridge collided with the offshore Mesozoic and Cenozoic trench leaving Antarctic devoid of plate boundaries, largely aseismic and with very few active volcanoes.

West Antarctic Rift System

Whilst Antarctica has been left isolated at the South Pole, surrounded by the Southern Ocean, as the original core of Gondwana, Antarctica itself continued to rift. Marie Byrd Land moved away from East Antarctica forming what we now call the West Antarctic Rift System (WARS). The continental crust underlying the rift was stretched and thinned, the Transantarctic Mountains were uplifted on the flanks of the rift to form the spectacular mountain chain that we see today and there was associated volcanism. Volcanism in Marie Byrd Land began 36 million years ago and explosive eruptions continued until as recently as 7500 years ago. Eighteen large

Figure 2.26 Mt Hampton, one of the many volcanic cones in Marie Byrd Land. (Credit: John Smellie)

central vent volcanoes and 30 small satellite volcanic centres formed in Marie Byrd Land on the flank of the rift system, rising to elevations between 2 and 4000 m above sea level. Most of these impressive volcanoes are covered in snow and their bases are buried beneath the West Antarctic ice sheet. Mt Sidley is the tallest volcano in Antarctica at 4181 m high, rising 2200 m above the surrounding ice level. Only Mt Berlin is considered active, as it has a streaming fumarolic ice tower and an underlying ice cave with elevated air temperatures. Although mostly inactive today their characteristic volcanic profile can still be seen through the ice. A second volcanic field, the McMurdo volcanic province, formed on the opposite flanks of the rift system in the western Ross Sea region. Most volcanism occurred on or along the front of the Transantarctic Mountains and includes the well known large volcanic cones (Mt Melbourne, Mt Discovery, Mt Morning) mostly older than 5 million years, and the three major volcanoes that coalesced to form Ross Island (Mounts Terror, Bird and Erebus) building up over the past 5 million years. Mt Erebus remains as the most active volcano within this mostly extinct province, although some ancient volcanoes within this province show evidence of geothermal activity. Mt Melbourne has warm ground that steams near its summit, and an ash layer exposed in ice on the volcano's flank that suggests it erupted less than 200 years ago. Fumaroles also occur on a recently recognised active volcano Mt Rittmann.

The maximum topographic relief from the top of the flanking mountains to the subglacial basins within the rift system is approximately 7 km. The thinned and extended crust has subsided to allow thick sequences of sedimentary rocks to accumulate in marine basins and record an amazing unique record of how Antarctica has cooled from a 'greenhouse world' 35 million years ago with greenhouse gases very much higher than today to an 'ice house world' with its thick ice sheets. Scientist have drilled through this sedimentary pile to depths of over 1100 m to sample the rocks that record the advance and retreat of the Antarctic ice sheets through time. We have not yet sampled rocks that record the greenhouse–icehouse transition. This remains a key target that may enhance our knowledge of the future climatic conditions that could result in the collapse and ultimate disintegration of the Antarctic ice sheets.

Figure 2.27 Ice towers formed over gas outlets on the side of Mt Erebus, an active volcano. (Credit: Antarctica New Zealand)

Changing greenhouse–icehouse worlds

During the time period 210 to 140 million years ago, climate was uniformly warm to hot and wet worldwide, with a luxuriant cosmopolitan flora of conifers, cycads, ferns, tree ferns, ginkgos and herbaceous lycopods and horsetails. Some of the most spectacular fossil forests are preserved within 100 million year old rocks on the Antarctic Peninsula. Hundreds of fossil trees and shrubs are preserved in their growth positions buried within sandstones that were deposited during catastrophic floods from volcanic uplands. At this time, the trees were growing in latitudes as high as 70°S. It is thought that the plants tolerated the extreme polar light regime by becoming dormant in the long dark winters and flourishing during the summer days of midnight sun. The first appearance of fossils of flowering plants (angiosperms) 80 million years ago marked a change in Antarctic vegetation. Many fossil leaves are similar to those that live in subtropical climates today, indicating that warm wet climates extended southward to cover Antarctica in much warmer

Figure 2.28 Artistic recreation of what a Cretaceous forest on Antarctica might have looked like. (Credit: Robert Nicholls)

global climates. A mean annual temperature of about 19°C prevailed with warm frost-free winters. Subtropical flora grew in warm climates on the volcanic arc surviving the environmental catastrophe that saw the extinction of the dinosaurs 65 million years ago.

When the first angiosperms (flowering plants) evolved, the continuity of the landmasses enabled them to spread globally. As the continents moved apart and linkages were severed, descendants of the first angiosperms could spread only where migration routes remained open. Distinct modern floras of the various continents are products of evolution in isolation since the separation of land masses. Sixty million years ago, Australia's last Gondwana links were severed and sea floor spreading in the Tasman Sea isolated New Zealand. Evolution in Australia, New Zealand and South America was from the animals and plants of Gondwana, and the individual modern flora and fauna of all three clearly show these common roots. Ghosts of the Gondwana flora are common in present New Zealand vegetation, e.g. totora, kauri and *Nothofagus* (beech) trees.

Younger fossil plants (55–40 my old) hold vital clues to a major change in climate at that time. The warm loving types disappeared, replaced instead by trees that were tolerant of cool climates. The southern beech, *Nothofagus*, became most common in

Figure 2.29 Fossil tree trunk from Cretaceous sedimentary rocks on Alexander Island. (Credit: Bryan Storey)

Figure 2.30 *Nothofagus* fossil leaf from Sirius Formation. (Credit: David Cantrill)

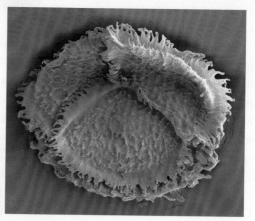

Figure 2.31 Fossil *Isoetes* spore 13 million years old from recently discovered lake horizon in the Dry Valleys. (Credit: David Cantrill)

Antarctic forests, along with the monkey puzzle conifer, *Araucaria*. Mean annual temperatures had dropped to around 10°C with winter temperatures below freezing. The onset of glaciation and cold climates led to the demise of Antarctic vegetation. Plant macrofossils are rare from rocks younger than 40 million years old but fossil pollen shows that tundra vegetation was present with small bushes of *Nothofagus*, mosses and rare conifers. In the Beardmore Glacier region of the Transantarctic Mountains, a unique flora of dwarf *Nothofagus* bushes, cushion plants and mosses is preserved sandwiched between glacial deposits. Preserved in their growth position, these plants grew only 300 miles from the South Pole and indicate a short burst of climatic warmth and glacial retreat during the icehouse world. These fossils heralded the end of Antarctica's ancient forest as the glacial landscape took over. Another amazing fossil horizon was discovered within the Victoria Land Dry Valleys in 2000. Soft organic material has turned out to be desiccated slabs of an ancient 14 million year old lake bed that has preserved freeze dried aquatic mosses, pea-sized freshwater ostracods, thick deposits of diatoms, pollen leaves and twigs of southern beech, aquatic quillworth plants and fragments of insects including a species of weevil. These fossils paint a picture of an alpine lake damned behind glacial moraines, surrounded by tundra and weather beaten southern beech shrubs 14 million years ago with temperatures at least 20°C warmer than today. It is very likely that this part of Antarctica has remained cold since this time as if this deposit had become warmer and wetter, microbes would have mined these deposits as carbon sources.

A changing world

Geological exploration in Antarctica has provided a unique record of how our Earth System functions, how it has evolved through time and how the continents amalgamated and combined in different configurations through geological time. The unusual position that Antarctica occupied as the central keystone of Gondwana enabled Antarctic geologists to provide crucial evidence in support of the theory of plate tectonics, to contribute to global debates surrounding the breakup and disintegration of large continents, to understand how our climate has changed and how plants and animals have evolved as the continents drifted through different climatic belts. Temporary land bridges influenced the migration and evolution of new species and constrained the global distribution of the Earth's species. Our investigations of the Antarctic rock record have allowed us to investigate and understand the interaction between the Earth's biosphere, atmosphere and lithosphere and how those interactions have evolved and functioned through time. The data provide a valuable benchmark against which we can monitor and predict future changes due to human-induced increases in atmospheric greenhouse gases.

The unusual combination of a continent that has undergone extension and stretching during the last 100 million years at a time when Antarctica plunged into the icy wilderness that we know today has resulted in a unique geological laboratory to test and model the future effects of climate change. We are faced with a situation where future global warming will accelerate the natural rates of change (cooling) that are currently being documented from the rock record of Antarctica through international drilling programmes and we may rapidly return to what Antarctica was like many millions of years ago.

Like other Gondwana continents, Antarctica is likely to have its fair share of mineral resources, glimpses of which have been recorded by Antarctic geologists as they systematically visited and mapped accessible rock outcrops over the past 100 years. The combination of known offshore sedimentary strata and organic carbon sources, as indicated by the abundant fossil flora, makes Antarctica a potential target for future oil exploration. Fortunately, it is currently not financially viable to extract oil and mineral resources from Antarctica, and the Madrid Protocol to the Antarctic Treaty prohibits mineral exploration for the time being. However, as the world's oil and mineral reserves diminish, we may see increased pressure to explore the potential mineral wealth of Antarctica. We can only hope that the world's nations will become less dependent on natural irreplaceable resources and that Antarctica remains protected for future generations.

Future challenges

Our future challenges as geoscientists is to continue to understand more fully the link between geological processes and changes in the Earth System particularly climate change. Antarctica witnessed dramatic changes over the past 50 million years from a continent that was a warm greenhouse world attached to South America to one that was transformed to an icehouse world isolated in a South Polar position. The rich flora and fauna of 50 million years ago were unable to live in the extreme conditions and were replaced by the plants and animals that could adapt to current conditions. We have much to learn about how these plants and animals adapted, how they might survive future warming, and what influence major physical changes like the opening of Drake Passage, and the uplift of the Transantarctic Mountains had on the climate system. We are very fortunate that Antarctica preserves a record of these changes that will enable us in the future to more fully understand how our Earth System functions and how it will respond to predicted changes in atmospheric composition and inevitable global warming.

3 Ice with everything

VALÉRIE MASSON-DELMOTTE

Where does the strange attraction to the polar regions lie, so
powerful, so gripping that on one's return from them one forgets
all the weariness of body and soul and dreams only of going back?
Where does the incredible charm of these unattended and terrific
regions lie?   **Jean-Baptiste Charcot, 1908**

For many people the word Antarctica is synonymous with cold and with
more ice than can be imagined. The Antarctic continent has been continuously
covered by a thick layer of ice for the past 15 million years, leaving less than 0.5% of
the underlying rock visible. This enormous mass of ice, formed by the progressive
accumulation of snowfall, slowly flows back to the ocean. Antarctica is a major
actor in our global climate system as well as being a most precious archive of
the history of climate evolution over the last 1 million years.

Antarctic ice in the global water cycle

For many millions of years, Antarctica has been the largest reservoir of
continental ice on earth. The dimensions of this 'sleeping giant' are astonishing.
The area permanently covered by continental snow and ice is greater than 13 million
km², 30% more than all of Europe. Around Antarctica, the cold Austral Ocean
enhances the ice area every winter as the frozen surface waters form sea ice. These
ice surfaces reflect sunshine back into space rather than absorbing it, helping to
maintain extremely cold conditions at the high southern latitudes.

The Antarctic sea ice is marked by spectacular seasonal variations. When
the sea surface temperature drops below −1.8°C, it can freeze. However, winds and
changes in ocean water density can induce a vertical mixing which increases the

Antarctica: Global Science from a Frozen Continent, ed. David W. H. Walton. Published by Cambridge University
Press. © Cambridge University Press 2013.

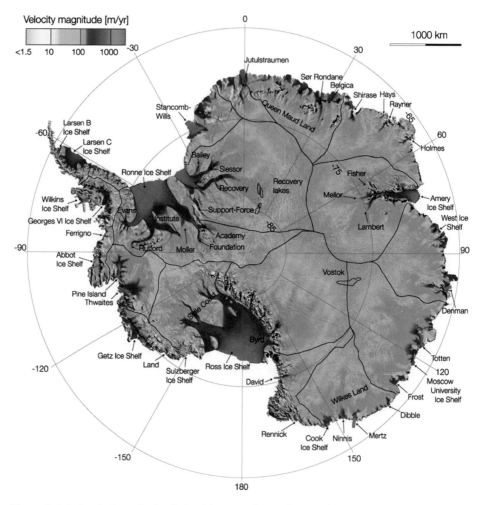

Figure 3.1 Antarctic ice velocity derived from satellite radar interferometry. Thick black lines delineate major ice divides. Thin black lines outline subglacial lakes. Thick black lines along the coast are ice sheet grounding lines. (Credit: Rignot, E., J. Mouginot and B. Scheuchl (2011). Ice flow of the Antarctic ice sheet. *Science*, **333**, 1427–1430)

temperature and thus delays the formation of sea ice. The initial ice is called frazil ice and its formation usually takes place in bays where the effect of wind is limited. As the air temperature continues dropping below −20°C, 'fast' sea ice forms *in situ*. This 2.5 cm thick soft ice is fractured by the tidal movements: when the tide rises, the expansion of the ocean surface creates ice cracks; when the tide falls, the decrease of the ocean surface generates an overlap of the slabs of ice at the edges. Near the shores, pancake-ice is formed when slabs of ice are jostled by the winds and waves and curl upwards at the edges. They can reach a thickness of 0.50 m.

Figure 3.2 Surface (top) and bedrock (bottom) topography of Antarctica. (a) Thin black lines correspond to elevations of 0, 500, 1000, 2000, 3000 and 4000 m above sea level. (b) Thin black lines represent modern sea level. The East and West Antarctic ice sheets are clearly visible. The bedrock of the East Antarctic ice sheet lies above sea level, making it a continental ice sheet. By contrast, most of the West Antarctic bedrock lies below modern sea level, and large parts of west Antarctica float as ice shelves at the ocean surface. A small retreat of such a marine-based ice sheet may destabilise the entire West Antarctic ice sheet. (Credit: Gaël Durand, using the data synthesis of Anne Le Brocq)

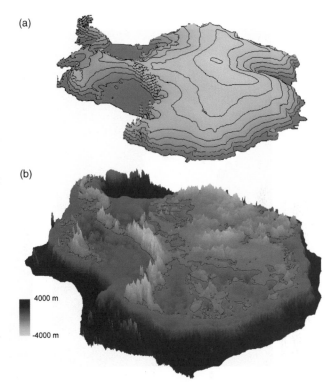

(a)

(b)

4000 m

-4000 m

In austral autumn and winter, the formation of sea ice in the Weddell and Ross sea sectors is intense and expands the sea ice area from 3 to 18 million km². The total area covered by ice is therefore restricted to the Antarctic continent (ice sheet), and the floating part of the Antarctic ice sheet (ice shelves) in summer, but more than doubles in winter when the Antarctic continent is surrounded by sea ice.

When sea ice starts forming, it is pushed northward by the intense Antarctic katabatic winds, blowing from the interior of the continent towards the sea. These winds induce the formation of coastal polynyas, which are areas of open ocean between the coast and the pack ice. These open waters undergo an intense cooling and produce fast ice. At the end of the winter, the stable winter sea ice breaks up and drifts away. Most of the winter sea ice melts and disappears completely. Because most of the Antarctic sea ice is driven northwards by the winds and currents away from the initial freezing area, it does not accumulate over many years and its thickness is usually only 1 to 2 m. On its journey northwards, the continuous cover of the Antarctic sea ice becomes fractured, leading to the formation of leads or cracks. Sea ice can also become covered by a thick snow layer, sometimes to the point where its weight pushes the sea ice down below sea level. The presence of an insulating snow layer will slow down the cooling and thickening of sea ice. The physical properties of the snow can also influence the sea

(a) (b)

(c) (d)

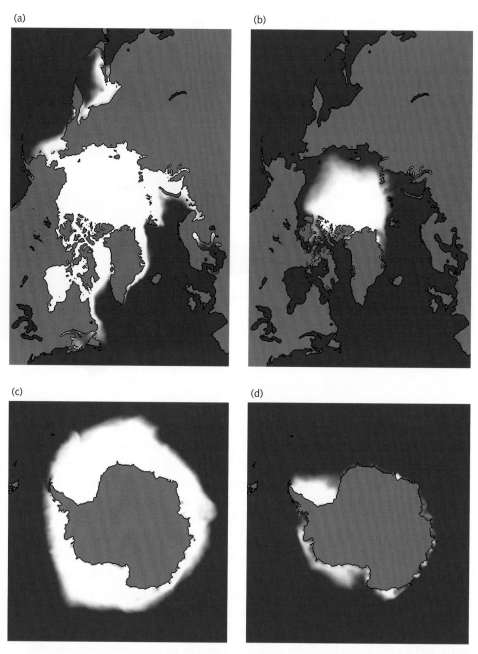

Figure 3.3 Maps of the Arctic (a and b) and the Antarctic (c and d) showing the winter and summer extent of sea ice. (Credit: Julienne Stroeve)

Figure 3.4 Pancakes forming at the ocean surface. Frozen seawater is shaped by the winds and waves. (Credit: IPEV/ M. Dufour)

Figure 3.5 The *Astrolabe* ship blocked in sea ice close to the Antarctic continent. (Credit: CNRS/ Erwan Amice)

ice albedo (its ability to reflect light and heat from the sun), its energy budget, as well as the rate of its expansion and decay. The presence of sea ice also has a large impact on the exchange of heat and water between the ocean and the atmosphere. Due to the huge winter sea ice cover, the 'continentality' and isolation of Antarctica is strongly enhanced in the winter: the distance between the open ocean waters and the Antarctic coast can increase by up to 3000 km!

Today, the extent of the Antarctic sea ice varies between approximately 2 million km^2 in local autumn (March) to 15 million km^2 in local spring (October) (Figure 3.3). A few reconstructions of past sea ice extent have been performed, based on biological indicators measured in deep sea sediment cores drilled near the modern sea ice margin. A synthesis of the data spanning the Last Glacial Maximum (about 21 000 years ago) suggests a doubling of the maximum sea ice extent, but limited changes in summer sea ice extent. Major efforts are needed to better document the history of the Antarctic sea ice extent, poorly known prior to satellite measurements.

Large inter-annual to decadal variability has been observed since the beginning of satellite measurements, in 1979. It has been observed that El Niño events, by heating the surface waters of the Southern Ocean, strongly affect inter-annual variations of Antarctic sea ice. Since 1979, the Southern Ocean has been warming at a rate of about 0.17°C per decade, but Antarctic sea ice shows a small increasing trend. Modelling studies suggest that two processes may be involved. First, the depletion of the Antarctic ozone layer has caused a cooling of the stratosphere, enhancing the westerly winds around Antarctica, strengthening the formation of coastal polynyas and sea ice production. Another explanation lies in changes in the mixing of surface ocean waters. An increase in precipitation amounts at the surface of the Southern Ocean, associated with warmer temperatures, can form a fresh, low density, surface ocean layer. This can reduce the mixing between the cold surface waters and the warm subsurface ocean waters, limiting the inflow of heat, which can melt sea ice.

This highlights the complexity of the mechanisms controlling the Antarctic sea ice, which remains a major scientific frontier. Intensive efforts have been deployed during the International Polar Year to better document the physical and mechanical processes associated with the sea ice formation, growth and decay. A radar altimeter boarded on the CRYOSAT-2 satellite is mapping precisely the thickness of sea ice. The improved knowledge from field and spatial observations

Figure 3.6 (opposite) In spring, sunlight and warmer temperatures induce the start of sea ice collapse. (Credit: IPEV/ D. Ruche)

Box 3.1 **Albedos of different types of snow and ice surfaces**

The albedo (from the Latin word 'albus', white) or reflectivity of a surface is a measurement of its efficiency in reflecting sunlight. Albedo is defined as the ratio of diffusely reflected to incident radiation.

The albedo of growing sea ice depends on its thickness, increasing strongly from values of 10% for 0–5 cm shallow sea ice to 35% for 30 cm thick sea ice. It also increases when sea ice cover breaks into small pieces. The presence of a thin layer of snow (a few centimetres) can also completely change the albedo, reaching values above 80% for dry and fresh snow.

The albedo of sea ice is one of the well-known positive feedbacks associated with climate change. In response to an initial change, a temperature increase can result in sea ice melting. The disappearance of sea ice, in turn, reduces the surface albedo and increases the absorption of sunlight, which increases the surface warming and the melting of the ice.

The albedo feedback is one of the mechanisms involved in the polar amplification of climate change, together with dynamic changes in atmospheric circulation, water vapour and cloud feedbacks.

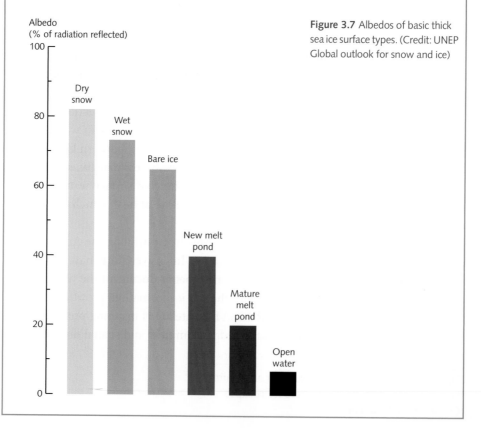

Albedo
(% of radiation reflected)

Figure 3.7 Albedos of basic thick sea ice surface types. (Credit: UNEP Global outlook for snow and ice)

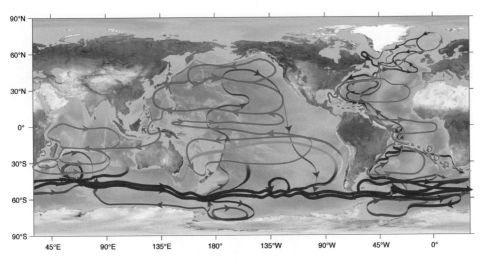

Figure 3.8 Schematic of the global ocean overturning circulation. Red indicates upper ocean flow, blue and purple are deep flows, and yellows and greens represent transitions between depths. (Credit: Sabrina Speich)

is crucial for an improved modelling of sea ice in global climate models. What is particularly challenging is the interplay between the small-scale features of the sea ice structure, its deformation, brine channels, which offer a major seasonal habitat for a wide range of organisms, and the large scale of its impacts on the Southern Ocean biology and climate.

The formation of sea ice in coastal polynyas has a dramatic effect on ocean circulation. Three processes are at play: first, the surface ocean water cooling enhances its density which is dependent on temperature. Second, as sea ice crystallises salt is squeezed out into the surrounding ocean, increasing its salinity and density. Finally, the freezing of ocean water removes heat from the ocean, due to the heat energy required for freezing. Together, these processes produce 10 million km^3 of the coldest, saltiest and densest ocean water in the world. This extremely dense water sinks towards the depth of the Southern Ocean and fills the deepest parts of the global ocean system, during a journey lasting thousands of years. These deep ocean masses play an important role in the global carbon cycle. The solubility of carbon dioxide in water increases when temperature decreases. As a result, the formation of deep-water masses near Antarctica transfers carbon dioxide from the atmosphere into the depths of ocean basins. Antarctica is not only the cold point of our climate machine, but also the factory for dense bottom ocean waters and plays a key role in carbon sinks. Therefore, what happens in Antarctica has direct relevance for our global climate.

Figure 3.9 Spring (November) evening light on the last sea ice fragments, a few Adélie penguins and the Astrolabe glacier. (Credit: CNRS Photothèque/Bruno Jourdain)

Figure 3.10 An Antarctic snowflake. (Credit: Rob Webster)

Antarctica is the highest, coldest and windiest place on Earth. The Antarctic ice sheet has an average thickness of around 1800 m, its maximum ice depth reaching more than 4770 m. On the coasts, thick platforms of ice flow onto the ocean surface, forming the world's largest ice shelves. They can reach hundreds of thousands of kilometres square in the Weddell Sea (the Ronne–Filchner ice shelf) or in the Ross Sea (the Ross Ice Shelf). The thickness of these ice shelves ranges from ~100 to ~1000 m. The total volume of the ice sheet is estimated at 30 million km^3, representing 80% of terrestrial freshwater. This amount of ice exerts an enormous pressure on the continental crust, pushing it down. To the east, the large continental ice sheet mostly stands above sea level. By contrast, the smaller West Antarctic ice sheet lies below sea level, and is therefore called a maritime ice sheet. In order to visualise the global significance of the Antarctic ice sheet, one can calculate the global sea-level rise that would

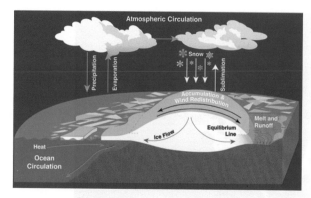

Figure 3.11 Schematic view of the interplay between an ice sheet and the global water cycle. (Credit: NASA)

be caused if all the continental ice was to melt. Altogether, the Antarctic ice sheet represents an equivalent of around 57 m of sea level, out of which 90% is for the continental East Antarctic ice sheet and about 10% for the maritime and more reactive West Antarctic ice sheet. Collapse of the marine unstable portions of the West Antarctic ice sheet would cause a 3.3 m global sea level rise.

The Antarctic ice sheet is fed by snowfall. The modern snowfall amounts vary strongly between the coastal areas which intercept cyclonic activity and intense snowfall (reaching sometimes several metres of water equivalent per year, in places such as Law Dome), and the inland dry deserts, with an annual mean snowfall amount less than 2 cm of water equivalent per year in places such as Dome A. On average, the snowfall on the Antarctic grounded ice sheet is estimated to reach 18 cm of water equivalent per year: each year, about 6 mm of global sea level are deposited as snow on the surface of the Antarctic continent. If the ice sheet was in equilibrium, then the amount of water returned to the ocean would be similar. Estimating changes in this balance of water in and water out is crucial to predicting the future contribution that the Antarctic could make to changes in world sea level but is very difficult to do accurately over such a huge area.

Day after day, year after year, snowfall accumulates as successive layers in the interior of the continent. Progressively, the ancient precipitation is buried in the ice sheet, writing silently the history book of the Earth's climate. The shape of the ice sheet results from the interplay between the inland accumulation of ice, the coastal melting, and the dynamics of ice flow initiated by gravity. Although it seems as hard as rock, ice needs to be imagined as a very viscous fluid that can flow downhill. Gigantic ice flow systems redistribute the ice within Antarctica, from inland to coastal locations. Unlike glaciers from other latitudes or Greenland, the major loss of mass from the Antarctic ice sheet is not caused by melt but results from the drainage of ice into the sea. Indeed, 90% of

Figure 3.12 Iceberg observed at 64°S from the *Marion Dufresne II* oceanographic cruise ship. (Credit: CNRS Photothèque 2003N00780/Xavier Crosta)

Figure 3.13 Small fragments of continental ice ranging from a few centimetres to a few tens of centimetres and transported by ocean currents off shore Dumont d'Urville until they melt. Heave and waves are very efficient in cracking icebergs and accelerating their fragmentation and melt. (Credit: CNRS/ Xavier Crosta)

the ice is lost from just 10% of the coast, especially through rapid ice streams: Amery, Ross and Ronne ice shelves, and those all around the Antarctic Peninsula. By contrast with the slow flowing interior (with a movement rate below 10 m per year), these streams can be rapid, with flow rates reaching 500 m per year, and their drainage basins extend over hundreds of kilometres square into the interior. After passing the so-called 'grounding line' at the coast, the floating ice further accelerates and gets thinner. The base melts into the sea and the dislocation of the ice shelves results in the calving of icebergs, finally bringing back to the ocean the water accumulated as ice in the interior.

When did the giant Antarctic ice sheet form? Three hundred million years ago, Antarctica was located far to the north in the Gondwana supercontinent, surrounded by a tropical ocean. Progressively, plate tectonics drove Antarctica towards its modern south polar location, reached about 80 million years ago. At the geological scale, the changing distribution of the continents altered the global carbon cycle which results from the interplay of carbon sinks (burial of organic matter and chemical weathering of silicate rocks which deposits carbonates on the

Figure 3.14 The Antarctic coast. This photograph highlights the impressive land ice. (Credit: IPEV/P. Le Mauguen)

ocean floor) and carbon sources (volcanic outgassing where carbonate-rich oceanic crust is subducted beneath moving continental plates). About 50 million years ago, the Indo-Asian collision produced the Tibetan Plateau uplift and a progressive loss of atmospheric carbon dioxide by weathering. Thus, the atmospheric carbon dioxide concentration declined over the past 50 million years and the climate cooled, with indications of initial Antarctic glaciation beginning more than 40 million years ago. When the Antarctic Peninsula was finally separated from the South American continent, about 34 million years ago, the Drake Passage isolated Antarctica from its closest neighbouring continent and opened the way for the establishment of the Antarctic Circumpolar Current, the world's largest current which continually drives the waters of the Southern Ocean around the Antarctic continent. It was then possible for the thick ice sheet on East Antarctica to expand, at the same time as West Antarctic glaciers grew, progressively building the huge continuous ice sheet we have today. The present East Antarctic ice sheet was probably formed 14 million years ago, while the West Antarctic ice sheet finished forming later (around 8 million years ago) when the Ross and Weddell Sea ice shelves could act as buttresses to support it. Despite probable minor variations in volume, the Antarctic ice sheet overall appears to have been rather stable over the past million years.

However, the matrix of the Antarctic ice sheet does not normally preserve ice that old. The renewal of ice is due to the vertical thinning and to the horizontal flow of ice towards the coastal glaciers. Due to this turnover between accumulation and flow, the average age of the ice preserved in East Antarctica is estimated to be around 125 000 years, while the modern West Antarctic ice sheet is estimated to be only 45 000 years old. Of course, these averaged numbers mask very large contrasts between coastal areas, where the turnover rate is large, and inland areas, where glaciologists still dream of drilling ice older than a million years at the stagnant ice domes.

Why drill in the ice?

Polar ice sheets contain unique archives of past climate and environmental changes. They provide information on local, regional and global climate, through the physical and chemical composition of the water and air preserved inside the ice matrix. Our knowledge of Antarctic climate based on instrumental records is frustratingly short, as monitoring instruments started being deployed only 50 years ago, during the 1957–58 International Geophysical Year. Obtaining past climate records is essential to characterise better the range of natural climate variability in Antarctica, from the inter-annual timescale to the past million years. Ice cores provide estimates of local climate parameters, such as the annual accumulation rate, a variable that is essential to characterise the mass balance of an ice sheet. In some low elevation coastal sites, summer melt occurs from time to time and this can be detected in the structure of the ice cores. In inland sites, where most ice core drilling operations have been conducted, temperatures are always too cold to allow surface melting. However, the initial snowfall still undergoes post-deposition effects due to wind erosion and sublimation/recondensation of water vapour in the porous snow surface.

The isotopic composition of the water trapped in snow and ice is the most valuable local climate indicator found in ice cores. Different stable forms of the water molecule are present on Earth: while most of the H_2O molecules are formed by two atoms of oxygen with 16 neutrons (^{16}O) and one atom of hydrogen, a small fraction of H_2O molecules are formed with 18 neutrons (^{18}O) (about 2000 parts per million of ocean water) or with deuterium ^2D (about 300 parts per million of ocean water). These different water molecules have the same chemical properties, but differ in their mass (due to the different numbers of neutrons) and also in their diffusivity (due to the different structures of the water molecules). In the atmospheric water cycle, water is evaporated at the ocean surface, cooled and therefore condensed in the atmosphere to form clouds, and finally precipitated above Antarctica to form snow. During each phase change, isotopic fractionation occurs. The moisture formed at evaporation is slightly impoverished in heavy molecules compared to the initial ocean water. Condensation also preferentially removes heavy molecules.

As a result, the transport of water vapour towards Antarctica is associated with a progressive distillation of water masses and a progressive loss of heavy water molecules. As early as the 1960s, a linear relationship was discovered linking the ratio of light to heavy water molecules of Antarctic snowfall with local temperature. About a thousand locations have been so far documented for their modern isotopic composition (see Figure 3.17), confirming its large-scale validity. This relationship, called the 'isotopic thermometer', is still commonly used to estimate past changes

Figure 3.15 Sastrugis formed by the erosion of surface snow by winds, on the traverse from Dumont d'Urville to Dome C. (Credit: IPEV/D. Ruche)

in temperature from mass spectrometry measurements of the light to heavy water molecule ratio in Antarctic ice cores. These data also provide a tool to test the realism of the water cycle simulated by atmospheric circulation models including the representation of the cycle of the different isotopic forms of water molecules. Such models have conversely been used to test the hypothesis that the present-day 'isotopic thermometer' applies for past climates.

The dating of deep ice cores also relies on stable isotopes. Glaciological models require assumptions on the relationships between stable isotopes in the ice, temperature and the accumulation rate (due to the saturation vapour pressure of the atmosphere). Glaciological age scales and independent age markers agree, supporting the use of stable isotope-derived temperature and accumulation estimates. Changes in snowfall patterns, deposition of 'diamond dust' and the strength of the temperature gradient up from the ground can all bias the relationship between snow isotopic composition and annual mean temperature. A conservative estimate is that past temperatures can be estimated using water stable isotopes with an accuracy of 20 to 30%, which represents an error of 2 to 3°C for the magnitude of glacial–interglacial changes, reaching around 10°C in central Antarctica.

Stable isotopes can also provide information on the evaporation conditions at the sea surface. Variations in deuterium excess are expected to be mostly controlled by the kinetic fractionation processes occurring both during evaporation and during snow crystal cloud formation. Thus precise measurements of both deuterium and oxygen 18 ratios of the same snow flake or the same piece of ice provide information on the evaporation conditions, in particular, the relative humidity at the ocean surface and the sea surface temperature. Data on deuterium excess from central Antarctic sites suggest a long distance, high altitude transport of moisture from the subtropics to inland Antarctica, while coastal locations receive moisture originating locally from the Southern Ocean. Ice core data therefore provide an integrated view of the water cycle, not only restricted to the deposition of snowfall inland on Antarctica, but also reflecting changes at the ocean surface much further north.

Together with snowfall, aerosols are deposited on the snow surface. These aerosols can be formed at the surface of our planet, through marine sea salt spray and salt flowers deposited on sea ice, by phytoplankton biogenic sulfur production, or uplift of continental dust all transported towards the polar atmosphere. Ice cores can therefore provide clues about environmental changes in remote areas, especially about sea ice extent or turbulence/windiness at the ocean surface. Detailed elementary and chemical analyses suggest that most Antarctic dust comes from Patagonia. Ice cores are thus truly an archive of many types of information: local

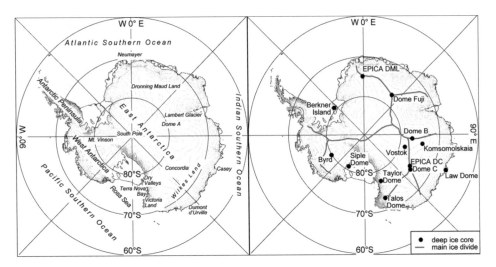

Figure 3.16 Map of Antarctica showing on the left, the main geographical sectors, and the right, the main deep drilling locations together with the main ice divides and the topography. (Credit: M. Frezzotti)

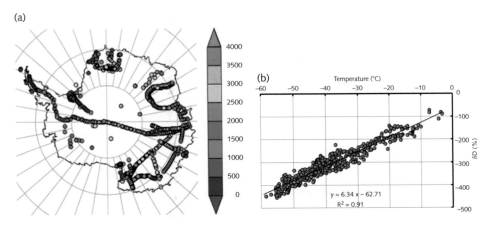

Figure 3.17 (a) Map of Antarctica showing the locations where snowfall, surface snow and deep ice cores have provided information on the local annual mean precipitation isotopic composition. Large areas have never been sampled. The colour scale reflects the elevation of the site (in metres above sea level). (b) Isotopic thermometer showing, from these locations, the relationship between precipitation, isotopic composition and temperature. A δD of 0 means that the sample has the same isotopic composition as the Standard Mean Ocean Water (SMOW). A δD value of −500‰ means that there are twice as few HDO molecules in the sample than in the SMOW, as observed in inland East Antarctica, near Dome A. The precipitation most markedly depleted in heavy water molecules is found in Antarctica.

Antarctic climatic conditions; climate changes in the sea ice and ocean moisture and salt source areas; aridity in South America.

The climate of our planet responds to perturbations of the global radiative budget. Natural factors, such as changes in the orbit of our planet, changes in the activity of the Sun, or injection of aerosols in the stratosphere by tropical explosive volcanoes can alter this radiative budget. Because such processes will force the climate to react, they are commonly called 'climate forcings'. While the orbital forcing can be accurately calculated from astronomical equations, specific aerosols archived in ice cores allow estimation of the variability of the solar and volcanic activity during the last millennia, which are indispensable inputs for climate models in order to simulate past natural climate variations. Both changes in solar activity and changes in the Earth's geomagnetic field can modulate the incoming cosmic rays at the top of the atmosphere. These cosmic rays produce a new isotope of beryllium (cosmogenic ^{10}Be), characterised by a million year long residence time. Measurements of ^{10}Be in ice cores have different applications: they provide a proxy for past variations in solar activity, which could play a significant role in decadal to centennial climate variability. Excursions or reversals of the Earth's magnetic field can be detected in Antarctic ice core ^{10}Be flux, which therefore provides a useful chronological marker for ice-core dating. Volcanic eruptions also act as an agent of climate change. Explosive tropical volcanoes distribute sulfate aerosols in the global stratosphere. These particles have both a direct effect on incoming solar radiation, reflecting it back into space, and an indirect effect as they act as cloud condensation nuclei. Elementary and chemical analyses of ice cores are used to characterise the magnitude and frequency of volcanic eruptions. On the scale of ice core records, volcanic activity appears randomly distributed in time. Climate reconstructions and climate models suggest that the small variations in solar activity and the frequency of volcanic eruptions are the main forcing agents for decadal to centennial climate pre-industrial fluctuations of the past millennia. Ice cores are therefore extremely precious as they provide information not only on the climate history, but also on some of the natural forces that act on climate fluctuations.

Finally, ice cores are the only direct archives of the past composition of our planet's atmosphere. The compaction of snowflakes into ice crystals entraps air bubbles. The analysis of the ice core air provides a direct archive of past changes in atmospheric composition. Present in small concentrations, some trace gases have huge impacts on the global climate system through their absorption of infra-red radiation (the so-called greenhouse effect). Ice cores reveal the natural variability of carbon dioxide (CO_2), methane (CH_4), and nitrous oxide (N_2O) concentrations back in time, and highlight the human imprint on the global atmosphere.

The ice-core scientific community has made great efforts to obtain the oldest possible ice core climate records. The formula is well known: you need to drill at a location where annual accumulation is as low as possible, where ice thickness is as high as possible, and where the ice flow is dominated by vertical thinning with limited horizontal movements. While the formula is accepted, its success requires a great deal of sophisticated logistic and technical knowledge. The first deep drilling operation conducted in West Antarctica, at Byrd, reached about 80 000 years back in time and characterised the climatic evolution over the past ice age. A long-term drilling effort at Vostok provided, in the mid-1990s, 400 000 years of climate and environmental evolution, together with the record of the deepest ice core ever drilled on ice sheets, to a depth of more than 3600 m below the snow surface. Recently, a consortium of 12 European laboratories from 10 nations has extracted the oldest available ice core record from Dome C, also located on the East Antarctic Plateau. Launched in 1996, the European Project for Ice Coring in Antarctica (EPICA) transported 1000 tonnes of material over a 1200 km traverse from the coastal station Dumont d'Urville to Dome C (Concordia Station), and eight drillers and 20 scientists worked in the field during the short summer seasons (8 to 10 weeks from November to January), year after year, from 1996 to 2005. A total depth of 3270 m was reached, spanning the past 800 000 years. This record duration was soon challenged by a Japanese drilling effort conducted at Dome Fuji, which has reached an almost comparable time span, 700 000 years back in time, and most recently a core of 3331 m has been extracted by the United States from the West Antarctic Ice Sheet, promising extraordinary high-resolution records of the last glacial climate.

Figure 3.18 displays some of the key results obtained from the analysis of the EPICA Dome C ice core over the past 800 000 years. The long-term climate evolution is marked by the succession of long glacial periods, characterised by an intense ice sheet growth in the northern hemisphere (and a decreased global sea level by up to 120 m), and relatively brief interglacial periods. The Antarctic temperature, derived from the analysis of stable isotopes of the EPICA Dome C ice core, varies in parallel with northern hemisphere glaciation. During ice ages, the magnitude of Antarctic cooling reaches ~10°C, twice the estimated global cooling of ~5°C. The long Dome C record also highlights distinct changes from one interglacial period to the next. Prior to 400 000 years ago, interglacial periods appear as 'lukewarm', compared to the intensity of the past four warmer periods.

The last interglacial period, occurring about 130 to 115 000 years ago, is marked in Antarctica, with temperatures 2 to 5°C above present-day levels. This past period is particularly interesting for understanding the reactions of the ice sheet to past climate changes. Indeed, emerged beaches and corals clearly demonstrate that the global sea level was 5.5 to 9 m above present-day, suggesting significant retreat of Greenland and/or West Antarctic ice sheets. Recently, the recovery of a north Greenland ice core extending back to the last interglacial has put strong

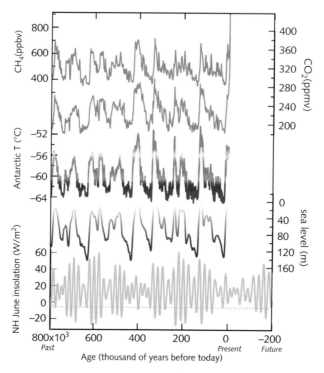

Figure 3.18 Past variations of, from top to bottom: green, methane concentration (parts per billion in volume, ppbv); grey, carbon dioxide concentration (parts per million in volume), derived from instrumental data and gas concentration measurements in EPICA Dome C ice core; EPICA Dome C temperature (°C) derived from stable isotope measurements of water. The colour shading is used to highlight the glacial periods (green to blue) and the interglacial periods (red colour for periods warmer than the current interglacial period); blue, global sea level change derived from deep ocean sediment Foraminifera isotopic composition. Note that cold periods in Antarctica coincide with northern hemisphere ice sheet growth and reduced sea level. The bottom curve (yellow) displays one component of past insolation changes, namely anomalies of mid-June insolation at 65°N (W m^{-2}). At this timescale, past changes in our planet's orbit around the Sun (its excentricity, obliquity and the position of the seasons) modulate the seasonal and latitudinal distribution of incoming solar radiation. Astronomical calculations of insolation are very precise for past and future changes, on timescales of tens of millions of years. The horizontal dashed line shows the insolation threshold required for the inception of an ice age. Such a threshold is not in view for the next tens of thousands of years in the future!

constraints on the Greenland contribution: in a context where the Arctic was also 3 to 5°C warmer, the northern part of the Greenland ice sheet was probably comparable to its modern extent. The southern part of the Greenland ice sheet was probably strongly reduced, with a contribution of about 2 m to the observed sea-level rise. This estimate suggests that the Antarctic ice sheet probably contributed to the observed sea-level rise of that period. The areas concerned may have involved the marine portions of the West Antarctic ice sheet as well as the Aurora/Wilkes Land sector of the East Antarctic ice sheet. However, there are still huge uncertainties on the sequence of events involved in the bipolar warming and ice sheet decay of the last interglacial period. Current deep drilling operations have recently been conducted in West Antarctica and in the Ross Sea sector, with the goal of improving the characterisation of the local climate and ice sheet evolution.

The Dome C climate record also shows interesting features of the relationships between the onset of ice ages and the variations of the Earth's orbit. The orbital theory of palaeoclimate relates the ice age changes with a latitudinal and seasonal redistribution of incoming solar radiation due to past changes in three features of the Earth's rotation around the Sun (excentricity, obliquity and precession of the equinoxes).

In this theory, the onset of an ice age is directly related to low levels of summer radiation received from the sun in the northern hemisphere. Indeed, both palaeoclimatic records and climate models show that such reduced summer insolation can prevent the melting of the snowfall accumulated during winter on the northern hemisphere high latitude continents. Climate feedbacks then amplify this to an initiation of ice sheet growth. Larger extents of snow increase the surface albedo and therefore the high latitude cooling effect. Increased Arctic sea ice extent and persistence, as well as changes in continental vegetation, amplify this high latitude albedo feedback, followed by changes in ocean circulation and in greenhouse gas concentrations, propagating the polar cooling towards the lower latitudes. A key question is then to identify which threshold of northern hemisphere solar insolation minimum is necessary to trigger the onset of an ice age. The Dome C record shows a particularly long interglacial period, about 400 000 years ago. Over the 20 to 30 000 year duration of that exceptionally long interglacial period, one small minimum of northern hemisphere summer insolation was not deep enough to provoke the onset of an ice age, while the second minimum was strong enough to trigger a glaciation. All other ice ages were triggered by stronger insolation minima.

What is the relevance of this finding? Future changes in the Earth's orbit can be calculated with high accuracy over millions of years. Over the next tens of thousands of years, the northern hemisphere incoming solar radiation will be exceptionally stable, due to the low excentricity of our orbit around the Sun, and will not cross this 'glaciation' threshold. This result suggests that, even without anthropogenic interference through greenhouse gas emissions, we are going to enjoy an exceptionally long interglacial period. In other words, we cannot count on the orbital forcing to compensate for the long-term warming induced by anthropogenic greenhouse release.

Box 3.2 **A slowly changing orbit**

The movement of our planet on its orbit is responsible for the distribution of incoming solar energy and the contrasts between low and high latitudes and seasons. If there were only the Sun and the Earth, the orbit of our planet would be a stable ellipse. However, the gravity of the other planets of the solar system interfere with the gravity of the Sun and induce slow changes in the orientation and shape of the orbit of the Earth. The eccentricity of the ellipse slightly varies over periods of around 100 000 and 400 000 years, changing the magnitude of the seasonal contrasts because of changing distances between the Earth and the Sun. The axis of rotation of the Earth has a varying tilt (called obliquity) with respect to the plane of this orbit. Obliquity, varying with periodicities around 40 000 years, modulates the gradient of insolation between low and high latitudes, and between seasons. Finally, movements of precession (hula-hoop) of the ellipse and of the axis of rotation of the Earth result in slow shifts of the position of the seasons on the orbit. The resulting climatic precession modulates the contrasts between seasons with a periodicity around 19 and 23 000 years. The orbital theory of palaeoclimates suggesting that periodic shifts in northern hemisphere summer insolation were driving ice ages was formulated in the 1940s by the Serbian mathematician Milankovitch. The importance of Earth's orbital parameters in driving past climate variations has been demonstrated since the 1970s through quantitative palaeoclimate reconstructions, and past climate simulations.

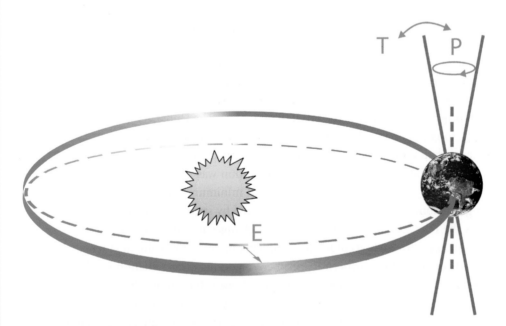

Figure 3.19 Schematic representation of the orbit of the Earth around the Sun, showing an exaggerated view of the excentricity of the orbit (E), the tilt of the axis of rotation (T), and the movement of precession (P). (Copyright IPCC 2007: WG1-AR4)

The Antarctic records of atmospheric composition are also bringing important insights into the coupling between the climate system and the global carbon cycle. Warm and cold Antarctic periods are associated with increased and reduced concentrations of carbon dioxide and methane in the atmosphere. Detailed measurements conducted on high-resolution ice cores suggest that during both the onset and the termination of ice ages, Antarctic temperature changes, driven by changes in insolation, occur just before or in phase with changes in atmospheric CO_2 concentration, which precede Nothern Hemisphere warming and methane concentration rise. At the glacial–interglacial scale, natural variations of carbon dioxide appear to be mostly controlled by changes in the ocean carbon uptake, with a key role for the Southern Ocean. Natural variations of methane, by contrast, are driven by methane emissions mostly provided by wetlands located in the tropics and boreal lands. Stable over the past 800 000 years, the apparent correlation between climate and greenhouse gas variations demonstrates the so-called carbon cycle/climate feedbacks. In response to an initial perturbation (here, orbital), climate change induces changes in the vegetation, soils and wetlands on the continents, and in the ocean temperature, gas solubility, circulation and sedimentation processes. These land surface and ocean changes in turn modify the sources and sinks of greenhouse gases. A climate warming induces an increased concentration of greenhouse gases in the atmosphere, itself inducing a supplementary absorption of infra-red radiation and a warming of the lower atmosphere and of the Earth's surface. Such a feedback is expected to continue to operate for the future: current anthropogenic emissions of greenhouse gases are inducing global warming, which is itself modifying the natural carbon sinks. Other climate feedbacks are revealed by the ice core data such as the amplifying role of changes in austral sea ice and Patagonian dust over the course of glacial–interglacial cycles.

Most climate models containing an explicit representation of the carbon cycle show that the doubts on future climate change arise from three key uncertainties: first, human emissions of greenhouse gases; second, the response of climate to increased greenhouse gas concentrations ('climate sensitivity'); and third, the carbon cycle feedbacks, with a saturation of the ocean and vegetation sinks leading to a larger proportion of the anthropogenic emissions staying in the atmosphere. The ice core data provide a benchmark to test the realism of these coupled climate and carbon cycle models, and provide observational constraints on climate sensitivity. Based on the magnitude of past Antarctic temperature variations, the available ice core data clearly support the prediction that a doubling of pre-industrial carbon dioxide concentrations in the atmosphere would induce a global warming of ~3°C.

Ice core data clearly show that the present-day levels of greenhouse gas concentrations are unprecedented over the past 800 000 years, and that the current rate of increase is unprecedented over the past 20 000 years, a time interval for which the resolution of ice core measurements is large enough to make this comparison. In fact, we can now say that the global atmospheric composition is at present controlled by the human emissions of greenhouse gases, which are well mixed throughout the atmosphere. The ice core perspective on the step change between the natural carbon cycle historical variability and the present industrial period is impressive, showing that we humans have become a 'geological force' at the planetary scale. With 30% more carbon dioxide and 200% more methane now than during any warm pre-industrial period, we have entered into a new geological period where humans are the dominant climate forcing factor: the 'Anthropocene'.

Emissions of greenhouse gases are already reaching 30 billion tonnes of carbon dioxide per year, and are increasing year on year. Future climate change scenarios have been explored by climate modelling groups, coordinated within the International Panel on Climate Change. Future global temperature change may vary between 1.5 and more than 4°C, depending on the emissions of greenhouse gases. Climate models consistently show a polar amplification of global temperature changes. In the central Antarctic plateau, the simulated temperature rise is about 20% higher than the global average. How would an Antarctic warming of a few degrees over the next century compare with the natural variability of Antarctic temperatures? There is evidence for interglacial periods when Antarctica (but not necessarily the whole planet) was up to 5°C warmer than at present. Therefore, a warming of several degrees above present day is not unprecedented. Past sea-level data show that the polar ice sheets were vulnerable to such warmth and that they contributed to an increased global sea-level rise by several metres. The fastest historical temperature rise we have detailed data for comes from the Dome C ice core and took place during the last deglaciation, rising at most 4°C within a thousand years. Worryingly, the future temperature rise simulated by present climate models could reach the same warming magnitude but about 10 times faster than at any time in the last 800 000 years.

Antarctic and Greenland ice cores do not only document the glacial–interglacial cycles linked with slow changes in the Earth's orbit. They also provide clear evidence for rapid climatic fluctuations operating over glacial periods. Greenland ice core data reveal 25 abrupt climate changes punctuating the last glacial period. These Greenland 'Dansgaard–Oeschger' events appear as prolonged cold periods, followed by an abrupt warming reaching up to 16°C in a few decades, and then a progressive cooling back to another cold episode. These rapid instabilities are not restricted to the high latitudes of the northern hemisphere, but they are also recorded in marine and terrestrial palaeoclimatic records at temperate and tropical locations. Interestingly, methane concentrations show rapid variations synchronous with the Greenland changes. The rapid variations in global methane were used to construct common age scales for both Greenland and Antarctic ice cores. Precise comparisons of Antarctic and Greenland temperature reconstructions have now shown that cold Greenland phases ('stadials') are associated with Antarctic warming. By contrast, the abrupt Greenland warming marks the end of the Antarctic warmth, followed by a slow decline. In this respect, the abrupt events have a global signature, but are not globally synchronous. The systematic bipolar see-saw behaviour confirmed that these abrupt climate events are associated with a redistribution of heat between the two hemispheres, caused by reorganisations of the global ocean circulation. Marine sediment palaeoceanographic records have clearly shown that the north Atlantic Ocean circulation has undergone major historical changes. These ocean circulation reorganisations were linked with massive freshwater discharges provided by the huge ice sheets which used to cover North America and north Eurasia. The formation of north Atlantic deep waters is indeed strongly sensitive to the density of surface waters, and a substantial inflow of freshwater can modify the density gradients, the formation of deep waters, and the efficiency of heat redistribution by the global ocean circulation. Recently, the long-term Antarctic methane and temperature record has revealed that these instabilities are not restricted to the last climatic cycle, but are a characteristic of climate change during all the glacial periods of the past 800 000 years. Past climates therefore show that unstable ice sheets can induce major reorganisations in global ocean circulation.

Do we need more ice cores? We need a better high-resolution description of the spatio-temporal climate and environmental changes in Antarctic. The recent history of Antarctic climate, mass balance, atmospheric aerosol load and climate forcing factors remains largely unknown, and the International Partnership for Ice Core Science aims to obtain high-resolution ice cores spanning the past 2000 years with annual resolution from all around the Antarctic. The processes involved in the rapid climate changes of the last ice age clearly involve ocean circulation changes, but also large-scale atmospheric circulation changes. This raises the need for circum-Antarctic high-resolution ice core records spanning the past 40 000 years, which can provide precise records of the sequence of regional climate changes, in relationship to local sea ice feedbacks both during the last rapid events and during the last glacial–interglacial transitions. Finally, while the 800 000 year long Dome C record has already brought a wealth of information and questions regarding glacial–interglacial cycles, it is a major challenge for the ice core community to obtain deep ice cores reaching back to 1.3 million years or more which may be possible in several places close to Vostok, Dome C, or Dome A where the ice is over 3000 m deep. This is perhaps the most challenging place in the world to work, high up on the Antarctic Plateau with a mean annual temperature of $-58°C$.

Indeed, palaeoceanographic data show a major climate reorganisation probably took place from 1.3 to 0.8 million years ago. Prior to the last million years, ice ages were less intense but more frequent, occurring with a spacing of around 40 000 years. Since 800 000 years ago, ice ages have been more intense, and longer, reaching a duration of ~100 000 years. This major climate transition remains essentially unexplained, and could be linked with a long-term decrease in atmospheric carbon dioxide concentration and different climate responses to orbital parameters. Ice core data could help to test these hypotheses, and to document key feedbacks between continental dust, Southern Ocean and sea ice variations, carbon cycle feedbacks and Antarctic climate. Field work conducted during the International Polar Year has improved the current knowledge of Antarctic accumulation rates, bedrock properties and ice thickness, key information to start planning new deep drilling operations.

The Antarctic ice sheet is the slowest changing component of the climate system due to the time constant involved in its growth and decay, and it has persisted over successive glacial and interglacial climate fluctuations. During glacial periods, the global sea level was up to 120 m below the present-day, due to the storage of water inside the two ice sheets which were covering the continents of the northern hemisphere, both in America (the Laurentide ice sheet) and in Eurasia (the Fennoscandian ice sheet). At the Last Glacial Maximum, about 21 000 years ago, these two ancient ice sheets had a volume of 80 million km^3, more than twice the volume of the modern Antarctic and Greenland ice sheets combined. At that time, Antarctica was about $10°C$ colder than nowadays. Because the sea level was lower than at present, the position of the West Antarctic grounding line was much further out to sea than at present, inducing a thickening of the inland West Antarctic ice sheet. This theory is supported by the discovery of submarine moraines and former grounding lines and streams on the continental shelf, about 100 m below modern sea level. By contrast with the maritime West Antarctic ice sheet, which is controlled by the rate of loss of ice into the sea, the continental East Antarctic ice sheet volume is driven by the inflow, i.e. the accumulation rates on the Antarctic plateau. During glacial times, the accumulation rate was strongly reduced on the Antarctic continent, restricting the ice volume of East Antarctica. Altogether, it is estimated that the Antarctic ice sheet contributed about 10–15 m to the 120 m of sea-level change between the last glacial maximum and today.

How does this climate history affect the modern Antarctic ice sheet? The present-day flow of ice within the Antarctic ice sheet is still responding to the changes produced by the last deglaciation. For instance, the distribution of

Figure 3.20 Photograph of an EPICA Dome C drill head and a deep ice core. The diameter of the ice core is about 10 cm. (Credit: Photothèque CNRS/L. Augustin)

Figure 3.21 Photograph of a thin slice of an ice core, taken at a depth of about 500 m. Air bubbles are clearly visible. These air bubbles are slowly trapped in the ice matrix at a depth of 80 to 100 m below the snow surface. At deeper depths and therefore higher pressures, air bubbles are no longer visible as air hydrates are formed. (Credit: BAS)

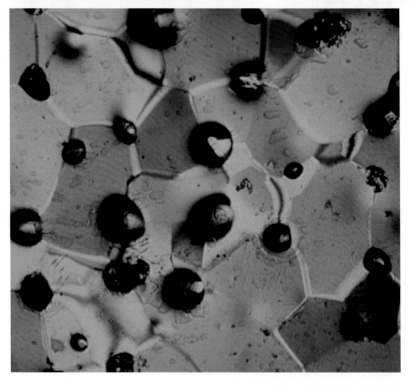

Figure 3.22 Photograph of a thin slice of ice from Law Dome, Antarctica, taken under polarised light with a microscope. The orientation of the ice crystals is reflected in their colour. The real length of the sample is about 1 cm. (Credit: Marc Delmotte, LSCE)

Box 3.3 **The slow trapping of air inside the ice matrix**

At the surface, the snow layer is porous and air can circulate within the snow. Due to the weight of the upper snow layers, snow is progressively transformed into ice. The densification process is associated with the formation of pores which are progressively closed and trap air in closed cells, stopping any further exchange with the atmosphere. At a depth of about 100 m (but depending on local conditions), air bubbles are formed. At a given depth, note that the trapped air is significantly younger than the surrounding ice: this age difference depends on the speed of the pore closure, which varies between about 20 years today for coastal sites with high snowfall up to several millennia in the driest inland sites such as Vostok or Dome A for glacial climates. Deeper in the ice, several hundreds of metres below the surface, air bubbles are no longer visible as air hydrates are formed and are trapped within the ice structure (clathrates). During the bubble enclosure process, the air components are affected by gravitational fractionation. In Antarctica, the comparison between air sampling from uncompacted snow, from ice core bubbles and from the atmosphere clearly shows that the air composition is reliably archived in ice cores.

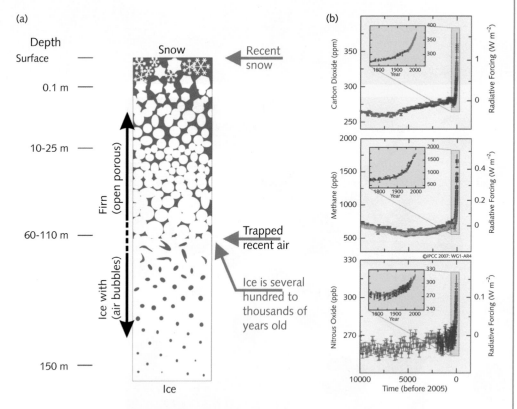

Figure 3.23 (a) Schematic view of the firn, the upper 100 m of the ice sheet, where snow is progressively compacted into ice, entrapping air into closed bubbles. (Credit: J. Schwander). (b) Comparison between greenhouse gas concentrations measured in Antarctic firn air and ice core bubbles (colour symbols) with direct instrumental measurements at South Pole (red). (Credit: IPCC, 2007)

Figure 3.24 Sunset on Concordia station, East Antarctica. (Credit: IPEV/J. Zaccharia)

temperature inside the central part of the ice sheet still preserves a signal linked with the last glacial climate. Similarly, the sub-Antarctic bedrock is still responding to the change in ice mass with time, a process called isostatic rebound where as the weight of ice decreases the bedrock gradually rises up. As a result, the West Antarctic grounding line may still be retreating and the ice sheet thinning, a multi-millennial process which could account for up to 0.5 mm of sea-level rise per year. The massive ice sheet reacts only slowly to climate perturbations by changing flow rates on timescales varying from a decade for fast ice streams to millennia for the inland ice.

Is Antarctica reacting to climate change?

While East Antarctic temperature appears rather stable, multi-decadal warming has been detected in the Southern Ocean waters as well as in West Antarctica, reaching 2.5°C in the Antarctic Peninsula. Until recently, there were not enough measurements to determine whether Antarctica was growing or shrinking. Different methodologies have been deployed to assess the mass balance of the Antarctic ice sheet. Over the past two decades, radar and laser altimeters on satellites have been used to monitor the surface height of the ice sheet and therefore estimate its volume. Over the past few years, new spatial instruments have begun to monitor the gravitational attraction of the Antarctic ice sheet and offer the potential to assess the interannual variability in its total mass. The mass balance change can also be analysed by comparing the input (snowfall accumulation) versus the output (water lost into the surrounding oceans). The input is estimated using ice core layer counting, meteorological measurements or atmospheric model simulations. These models can also be used to estimate the runoff caused by coastal melting. The mass loss due to calving processes is estimated from measurements of coastal ice flow velocity and thinning.

The different methods suggest that the West Antarctic ice sheet has been increasingly losing mass over the past years, at a rate of about 0.4 ± 0.2 mm per year, which represents 10–30% of the current global sea level rise (3.2 mm year^{-1}). With evidence that the overall Antarctic continent accumulation of snow has been stable over the past 50 years, the accelerating net loss of Antarctic ice appears mostly due to the outflow of ice from the Antarctic Peninsula and some West Antarctic

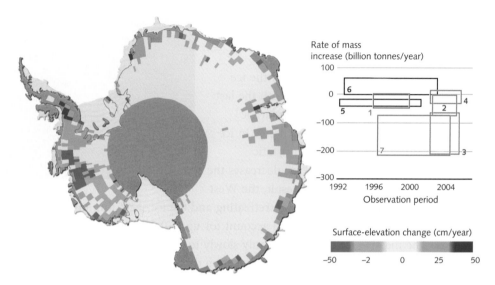

Figure 3.25 Rates of surface elevation change from 2002 to 2005 derived from satellite radar-altimeter measurements. (Credit: UNED/Global Outlook for Snow and Ice)

ice shelves. Satellite measurements have shown a rapid thinning of West Antarctic ice shelves, for instance in the Admundsen Sea sector, where the surface height has decreased by several metres over the last decade. The thinning of the ice shelves appears to be mostly caused by basal melting at the contact with relatively warm ocean waters lying below the shelves, and induces an acceleration of the tributary glaciers. The ice shelf reaction can be quite abrupt, as illustrated by the collapse of the Larsen B ice shelf in 2002, in the north-eastern Antarctic Peninsula. The analysis of ocean sediment cores from beneath the lost ice shelf has revealed that this collapse had not happened since the last ice age, and that it was associated with the acceleration of the tributary glaciers by factors of 2 to 8. Ice shelf retreat has also been recently monitored on the western side of the Antarctic Peninsula, with the collapse of the Wordie ice shelf in 2009 and the recent fracturing of the huge Wilkins ice shelf.

Fast ice sheet flow can be associated with positive, amplifying feedbacks. Typically, in response to air or ocean temperature rise, ice shelf melting or calving is expected to be associated with ice flow acceleration and thinning of inland ice. This thinning results in a lower elevation and a higher temperature, which in turn may increase the surface warming and melting, amplifying the initial decay. Recently, it has been shown that sedimentary processes are associated with movements of grounding lines. Till (ground up rock at the base of the ice sheet) delta deposited by ice streams may help to stabilise the position of the grounding lines, thicken the ice sheets, and lessen their vulnerability to

Figure 3.26 Photograph of an Antarctic Peninsula ice shelf near Cape Foyn prior to disintegration. Edge parallel crevasses indicate future calving. (Credit: NSIDC/ S. Tojeiro.)

small sea-level variations which could otherwise cause them to flex and break. If Antarctic surface air and ocean temperatures rise by several degrees in response to increased anthropogenic greenhouse gas concentrations, such warming may threaten the stability of the West Antarctic ice sheet, an issue highlighted as one of the key 'climate tipping points'. However, significant progress is needed to quantify the magnitude and timing of the Antarctic climate and ice sheet response risks.

Does Antarctica have wet feet? The evolution of the Antarctic ice sheet is also controlled by what happens at the interface between ice and the underlying rock. Surprisingly, an active subglacial water network has recently been identified beneath the Antarctic ice sheet. At the base of the ice sheet, the interplay between the dissipation of energy caused by ice friction as it flows over the bedrock and geothermal heating rising up from the core of the Earth induces 'warm' temperatures, corresponding to the ice melting point. The melting and freezing processes control the interaction between subglacial water and the overlying ice sheet.

Lake Vostok is an iconic image for such basal water; the analysis of the deepest part of the Vostok ice core has demonstrated that the bottom part of the ice sheet is locally formed by the refreezing of subglacial lake water. Most of the almost 200 subglacial lakes identified to date are located in the interior of the ice sheet, within 100 km of the ice divides and ice domes, and most of them are less than 20 km in length. Their volume is about a quarter of the total water in continental surface lakes. Satellite imagery has been used to identify the distribution of Antarctic subglacial water bodies, including dozens of lakes and rivers. Simultaneous rise and fall of surface elevations at several places on the Antarctic ice sheet surface have revealed that there is extensive water movement below the ice from one year to the next, and that the ice sheet morphology reacts to the drainage and flow of subglacial water. Such a large-scale drainage network is expected to modify the lubrication of ice sheet beds and the ice flow. Catastrophic drainage of subglacial lakes could be closely related to the onset of rapid ice stream flow. It is a challenge to characterise the subglacial Antarctic water systems and their dynamics. Data suggest that some of these lakes may have a turnover rate of several thousand years, up to hundreds of thousands of years for large lakes such as Lake Vostok. Catastrophic drainage of some lakes could have delivered large volumes of melt water to the Southern Oceans. The amount of water is not significant in terms of global sea level, but it may play an important role in triggering rapid ice flow and its associated contribution to sea level rise. Ocean modelling suggests that the formation of deep water in the Southern Ocean may be sensitive to large freshwater supplies, for instance in the Ross Embayment.

Antarctica could be the sleeping giant of our climate system. Lessons from the past have shown that high levels of atmospheric greenhouse gas composition were associated, a few tens of million years ago, with a planet free of continental ice. They have also revealed that northern hemisphere maritime ice sheets of the last glaciations have undergone surge instabilities associated with abrupt ocean circulation and climate changes. The most recent observations have, however, revealed that the giant could wake up faster than expected. The presence of an active hydraulic subglacial system suggests the potential for rapid changes in ice flow. Recent measurements suggest that Antarctica is now losing mass, mostly due to basal melt below West Antarctic ice shelves and increasing rates of ice flow down to the sea. Glaciologists face numerous challenges to monitor this huge mass of ice, to invent new methods to decipher the continental and marine subglacial processes, and to produce a new generation of ice sheet models able to resolve the ice streams and ice shelves.

Today, Antarctica has started to contribute to global sea-level rise. Have we wakened the sleeping giant? How much of the observed reactions are caused by anthropogenic interference with natural variability in global processes? What is

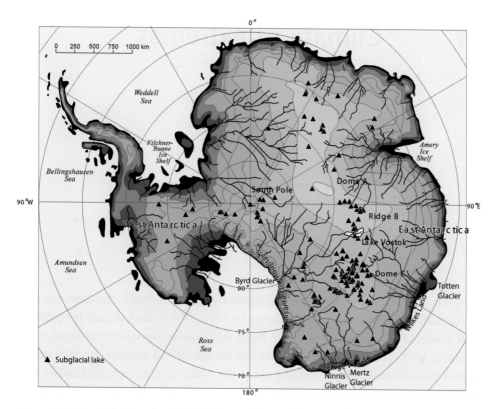

Figure 3.27 Identification of subglacial lake and drainage systems. (Credit: M. Siegert)

expected from the interplay between the local forcing due to ozone decay and the global forcing due to anthropogenic greenhouse gas emissions? What could be the contribution of Antarctica to sea-level rise over the next centuries? What should we expect in terms of changes to sea ice, deep ocean water formation and carbon sinks? How may these changes affect the unique and fragile Antarctic ecosystems? Intensive observational campaigns are required to improve our capability to describe, monitor and analyse the processes at play, and improve the numerical models used to assess the magnitude and pacing of future changes.

How much and how long can we shake the sleeping giant before waking it up?

4 | Climate of extremes

JOHN J. CASSANO

> We had discovered an accursed country. We had found the
> home of the blizzard.
> **Sir Douglas Mawson, Australasian Antarctic Expedition, 1911–14**

Antarctica is the coldest, windiest and driest continent on Earth. Temperatures at the surface of the continent have fallen to −89°C, the coldest surface temperature ever observed on Earth, and monthly mean wind speeds have been recorded in excess of hurricane strength. Antarctica can be considered a desert, yet 99% of the continent is covered in ice. What is it about the Antarctic continent that causes such extreme weather and how does Antarctic weather affect the rest of the planet?

Scientists have been fascinated by Antarctica's weather since the earliest explorers returned with unbelievable descriptions of storms, cold and wind. We now know much more about the processes that create this unique, harsh and dramatic climate and how this distant polar climate impacts those of us who live in more hospitable locations. While it may be difficult to see how Antarctica, a continent at the bottom of our planet and far distant from where most of us live, can impact our lives, Antarctica plays a central role in shaping and driving our global climate and the changes that occur there can have far reaching impacts on the rest of the planet.

The Earth's radiation budget

At the simplest level the climate and weather of Earth is driven by an imbalance in the amount of energy gained in the tropics and the amount of energy lost at the poles. We need to know where this energy comes from and why there is an

Antarctica: Global Science from a Frozen Continent, ed. David W. H. Walton. Published by Cambridge University Press. © Cambridge University Press 2013.

Figure 4.1 Cover of *The Home of the Blizzard* by Douglas Mawson, published in London in 1915. (Credit: D. W. H. Walton)

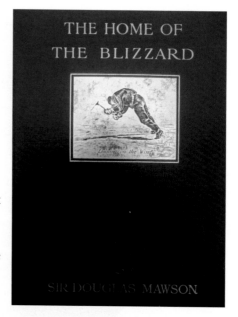

imbalance between different regions of our planet before we can understand the processes responsible for creating the unique climate of Antarctica.

Aside from a tiny fraction of energy gained from geothermal processes within the Earth all of the energy that drives the climate, and allows life to exist on our planet, comes from the Sun in the form of electromagnetic radiation. This radiation from the sun is often called solar radiation and includes visible light as well as forms of 'light' which we cannot see, such as ultraviolet (UV) radiation, which is responsible for causing sunburn and skin cancer. All objects on Earth, including the atmosphere, also emit radiation. The range of 'light' contained in this emitted radiation is referred to as longwave radiation and is not visible to our eyes. If we consider a location at the edge of the Earth's atmosphere, on the fringe of outer space, the only way that energy can be gained or lost by the atmosphere is by the gain of solar radiation or the loss of longwave radiation. Solar radiation, having travelled from the Sun through the vacuum of space, arrives at the top of the atmosphere. As this solar radiation passes through the atmosphere it may be reflected back out to space by clouds, dust, or even the surface of the Earth. The remainder of this solar radiation is either absorbed by the atmosphere or by the surface of the Earth. In either case, the atmosphere or the surface of the Earth will be warmed when it absorbs solar radiation. Similarly, longwave radiation emitted by the surface of the Earth, by clouds, and by the atmosphere can pass upward through the atmosphere and eventually exit the atmosphere to pass into the vacuum of space. It is the balance between the net amount of solar radiation entering the atmosphere versus the amount of longwave radiation leaving the atmosphere that determines whether a particular location on the planet will warm or cool over time. As we will see the circulation of the atmosphere serves to redistribute energy from areas with an excess of radiation gain to areas with a net loss of radiation.

Over an entire year the amount of solar radiation entering the atmosphere depends on the location. More solar radiation enters the atmosphere in the tropics than in the polar regions. This spatial variation in the amount of solar radiation reaching the top of the atmosphere has to do with the geometry of the Earth's

Figure 4.2 Satellite image of Earth in the visible spectrum to illustrate differences in albedo between clouds, sea, land and ice. (Credit: Matthew Lazzara. Space Science and Engineering Center, University of Wisconsin–Madison)

orbit around the sun and the fact that the Earth's axis, around which the Earth rotates, is tilted relative to the Earth's orbital plane. From this difference alone it is not surprising that the tropics are warm while the polar regions are cold.

Further accentuating this difference in incoming solar radiation is the albedo, or reflectivity of the surface. A satellite image of the Earth at visible wavelengths shows that some regions are bright while others appear dark. Bright regions are typically the result of extensive cloud cover or extensive snow and ice cover. Oceans are often the darkest regions, with non-snow-covered ground in between the dark oceans and the bright ice-covered regions. Since 99% of the Antarctic continent is covered by ice much of the incoming solar radiation which reaches the ground is immediately reflected back to space, and thus does not contribute to warming this part of the planet. Of the solar radiation that does reach the Earth's surface in the Antarctic, approximately 80% will be reflected back to space. This albedo effect then further enhances the energy, and thus temperature contrasts between the polar regions and the tropics.

There are seasonal variations in the amount of solar radiation reaching the top of the atmosphere. In the tropics, again as a result of the geometry of the Earth's orbit around the Sun, there is little variation in the amount of solar radiation entering the atmosphere from one season to the next. As a result the temperature in the tropics remains relatively constant throughout the year. Contrast that with a location such as the South Pole. Here solar radiation enters the top of the atmosphere continuously for 6 months of the year followed by 6 months when no solar radiation is received. This large difference in the amount of solar radiation entering the polar atmosphere between summer and winter results in a large seasonal temperature range.

In looking at the net energy gain or loss from the atmosphere we must also consider the longwave radiation lost from the top of the atmosphere. The amount of longwave radiation lost is determined by the temperature of the atmosphere, with warmer locations emitting more longwave radiation than cooler locations, so that more is lost in the tropics than in the polar regions. The difference in the amount of longwave radiation leaving the atmosphere between the tropics and the poles is much smaller than the difference in the amount of solar radiation entering the atmosphere between the tropics and the poles, so the differences in solar radiation dominate the energy differences between the poles and the tropics.

Considering these gains and losses we find that there is a net gain of energy in the tropics, which extends to approximately 37°N and 37°S. Poleward of these locations there is a net loss of energy by radiation at the top of the atmosphere. It therefore comes as no surprise that the tropics are warm and the polar regions are cold, but the gain of energy in lower latitudes year after year should result in these regions getting progressively warmer every year, as more and more energy enters this part of the atmosphere. Similarly, the high latitudes should become colder and colder with each passing year, as more and more energy is lost from these areas. But this is not what is observed, so what then can explain this apparent contradiction?

Winds in the atmosphere and currents in the ocean act to transport energy from warm regions to cold regions. In effect the circulation of the atmosphere through winds and storms and of the oceans by currents counters the imbalance in the radiation budget at the top of the atmosphere. These atmospheric and oceanic circulations exist because of the temperature gradient between the tropics and the poles, and ultimately work to remove this temperature difference. This temperature difference never disappears, as the radiation imbalance always drives

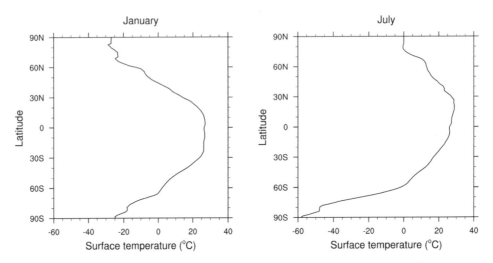

Figure 4.3 Latitudinal variation of surface temperature (January and July) averaged along each line of latitude. Note how the temperature difference between the poles and equator varies from January to July. This temperature difference is largest in the winter hemisphere (January in northern hemisphere and July in southern hemisphere). (Credit: Matthew Higgins, Cooperative Institute for Research in Environmental Sciences, University of Colorado)

the atmosphere back to a state in which the tropics are warm and the polar regions are cold.

From the above discussion we see that the Antarctic, with its year-round loss of energy from the top of the atmosphere, is critical in creating the pole to equator temperature gradient which drives both the atmospheric and oceanic circulations on our planet. Changes in the Antarctic continent, such as a loss of ice cover, would alter the energy balance over the continent, which would then alter the temperature gradient from the Antarctic to the tropics, and ultimately alter our global atmospheric and oceanic circulations. The effect of changes to the energy balance over the Antarctic could result in changes to the global thermohaline circulation in the ocean or could alter the path of storm systems in the mid-latitudes where many of us live. The weather and climate of our parts of the planet are thus intimately tied to the weather and climate of our most remote continent, Antarctica.

Global atmospheric circulation

The movement of warm and cold air in the atmosphere acts to reduce the temperature gradient between the tropics and the poles. If the pole to equator temperature gradient increases either through the tropics warming or the polar regions

Box 4.1 **The three-cell model of global circulation**

The pole to equator temperature gradient, combined with the fact that we live on a rotating planet, produces several circulation cells in the atmosphere. We can think of these cells as extending vertically through the lowest 10 to 18 km of the atmosphere, where temperature gradients are largest. This lowest layer of the atmosphere is known as the troposphere, and the top of this layer is known as the tropopause. In the tropics warm, humid air rises in thunderstorms through the lowest layer of the atmosphere. When this rising air reaches the tropopause, the boundary between the troposphere and the stratosphere, it stops rising and begins to move laterally towards both poles. This poleward moving air then begins to sink in the subtropics, near 30°N and 30°S. When this sinking air reaches the surface of The Earth some of it is deflected towards the equator, completing one cell of the global circulation – known as the Hadley cell. The remainder of this sinking air begins to flow towards the poles. In the mid-latitudes, near 50 or 60° latitude, this poleward moving air encounters equatorward moving air that originated in the polar regions. As these two airstreams collide they are forced to rise. This rising air ultimately reaches the tropopause and ceases to rise. Some of this air is deflected back towards the equator, but sinks near 30° latitude, completing a second circulation cell known as the Ferrel cell. The remainder of the air that had reached the tropopause near 50 or 60° latitudes begins moving towards the poles. This air sinks over the poles, completing the third and final cell of the global circulation, which is known as the polar cell.

The equatorward and poleward directed portions of these circulation cells will transport either cold polar air towards lower latitudes or warm tropical air towards the poles. Through this circulation the Antarctic 'exports' its cold air to the mid-latitudes of the southern hemisphere, as can be attested to by the residents of places such as Australia, New Zealand, southern South America and southern Africa, when cold southerly winds invade their homes and bring the possibility of snow.

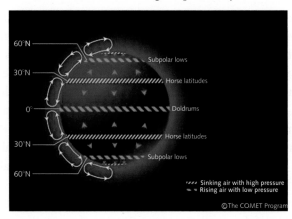

Figure 4.5 Schematic of the north–south and vertical winds that comprise the global circulation. This schematic is referred to as three-cell global circulation model since each hemisphere contains three circulation cells. (Credit: Elizabeth Lessard, Cooperative Program for Operational Meteorology, Education and Training (COMET), University Corporation for Atmospheric Research (UCAR))

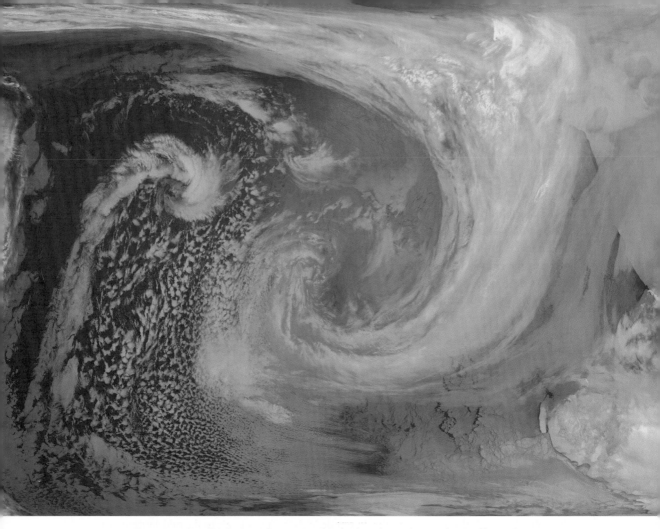

Figure 4.4 Satellite image of an area of polar low (storm) over the Southern Ocean. (Credit: NOAA)

cooling, this circulation would intensify. In fact, this intensification of the circulation is seen every year. Consider winter in the southern hemisphere. The Antarctic continent is in the middle of 6 months of darkness, while the tropics are warm. A large temperature contrast exists at this time of year since the difference in radiation gain between these regions is at a maximum. As a result the portion of the global circulation in the southern hemisphere is most intense at this time of year. This is seen in stronger storms and faster jet stream winds. In the summer, when the Antarctic is experiencing 6 months of daylight, the temperature contrast between the tropics and the polar regions is smaller, and as a result the circulation cells (see Box 4.1) in the southern hemisphere are weaker, as evidenced by weaker storms and a slower jet stream.

The rising and sinking portions of these circulation cells are also quite important. Through these vertical branches of the circulation warm air rises through the atmosphere while cold air sinks. As warm air rises through the atmosphere it cools, due to the decrease in pressure with increasing altitude and the resulting expansion of the air, similar to air that is let out of a tyre. As the air cools clouds, and eventually precipitation, can develop. In fact, most of the clouds and precipitation that occur in the atmosphere are the result of rising motion. Knowing this, it is not surprising that the region near the equator, associated with the rising branch of the Hadley cell, is quite cloudy with frequent thunderstorms. Similarly, the rising branch of the Ferrell and polar cells near 50 to 60° latitude are also characterised by extensive cloudiness and precipitation. As air sinks in the atmosphere it warms and this warming causes any clouds which may be present to dissipate, thus suppressing the development of precipitation. Not surprisingly, the regions near 30° and the poles, which are characterised by sinking motion between the Hadley and Ferrell cells and in the polar cell, have generally cloud free skies and receive little precipitation.

Solar radiation: warming the atmosphere

When solar radiation is absorbed by the ground the ground warms, in the same way that you will be warmed if you stand in bright sunlight. The majority of solar radiation that is absorbed by the Earth (i.e. the atmosphere and the ground) is absorbed by the ground. As a result the surface of the Earth is heated and in turn warms the air in contact with the ground, similar to the way a pot of water placed on a stove is warmed by contact with the hot burner. As one moves up through the atmosphere, away from the ground, the warming influence of the ground decreases, and as a result the temperature of the atmosphere also decreases.

This decrease of temperature with increasing altitude in the atmosphere is typical for the lowest layer of the atmosphere, known as the troposphere. This tendency for temperature to decrease with increasing altitude eventually ceases, usually at altitudes between 10 to 18 km above sea level. Above this height temperature either remains constant with altitude or actually increases with altitude. This situation, where temperature increases with altitude, is known as an inversion. The layer of the atmosphere above the troposphere in which temperature is either constant or increasing with altitude is known as the stratosphere. From the discussion above we see why temperature decreases with altitude in the troposphere, but why does the temperature increase with altitude in the stratosphere?

It turns out that a large fraction of the UV radiation that makes up solar radiation is absorbed by the gas ozone. The maximum concentration of ozone is found in the stratosphere. As a result, the stratosphere is heated by the absorption of UV solar radiation by ozone. The region of maximum absorption is in the upper stratosphere, and thus this is the region of the stratosphere with the warmest temperatures. As sunlight passes through the stratosphere, more and more of the UV solar radiation is absorbed, leaving less and less to continue down through the

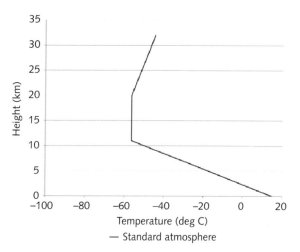

Figure 4.6 Standard atmosphere temperature profile showing the typical change of temperature with height in the mid-latitudes. The region of decreasing temperature with height between 0 and 11 km is referred to as the troposphere and the layer above with constant and then increasing temperature with height is referred to as the stratosphere.

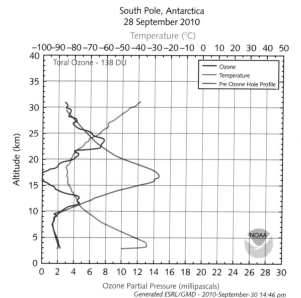

Figure 4.7 Vertical profile of ozone concentration over the South Pole on 28 September 2010 (shown by blue line). The green line shows the ozone profile found before the formation of the ozone hole. The red line shows the vertical profile of temperature. Notice that the lowest ozone values occur at heights with the coldest temperatures. (Credit: NOAA)

atmosphere. As a result the warming of the stratosphere, by absorption of UV solar radiation, decreases as one moves down through the stratosphere, resulting in cooler temperatures. This change in the amount of UV solar radiation absorbed by ozone with height in the stratosphere gives the stratosphere its characteristic increase in temperature with increasing altitude. The absorption of UV radiation by ozone in the stratosphere is also of great importance for life on Earth, as UV radiation is quite energetic and can cause damage to living organisms, resulting in sunburn and, in extreme cases, skin cancer and cataracts.

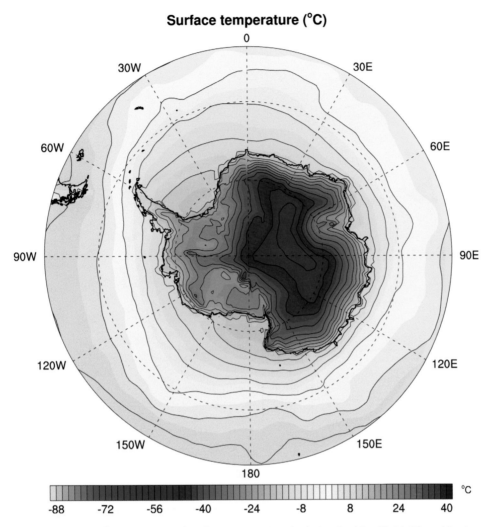

Figure 4.8 Map of average annual surface temperature in Antarctic. (Credit: Matthew Higgins, Cooperative Institute for Research in Environmental Sciences, University of Colorado)

The near surface temperature of the Antarctic: why is Antarctica so cold?

Since Antarctica is located in the polar regions there is a net radiation deficit in this region. This radiation deficit results in cold temperatures, but those cold temperatures are moderated by the transport of warm air towards the poles by the global circulation. Without this transport of warm air Antarctica would become colder with each passing year. Summer temperatures in the Antarctic range from just above 0°C near the coast to −30°C in the interior.

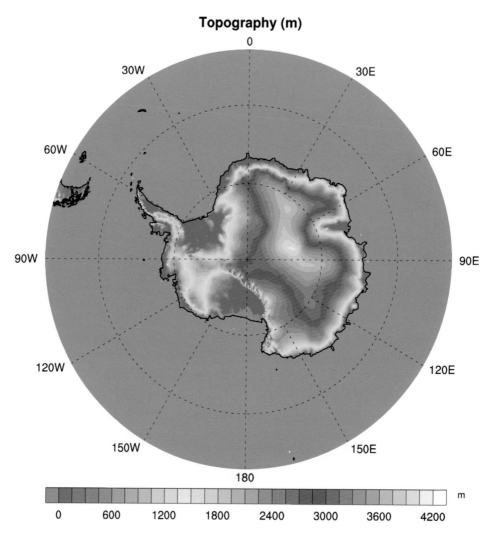

Topography (m)

Figure 4.9 Map of Antarctic topography showing the high East Antarctic plateau and lower West Antarctic ice sheet. Both regions slope steeply down to the coast. (Credit: Matthew Higgins, Cooperative Institute for Research in Environmental Sciences, University of Colorado)

In the winter temperatures near the coast are near -15°C while in the interior temperatures drop to –65°C.

Weather observations from the Antarctic indicate that the Antarctic is colder than the Arctic. The radiation deficit is identical in both polar regions, so why is Antarctica colder? The colder climate of Antarctica is due to the fact that Antarctica is an elevated ice sheet, with maximum elevations over 4000 m, while in the Arctic the North Pole is located at sea level. Since temperature decreases with elevation in the troposphere and the fact that Antarctica is at a much higher elevation than

Figure 4.10 Winter (red line) and summer (green line) temperature profiles over the South Pole compared with the typical temperature profile found in mid-latitudes. The winter profile shows a strong surface inversion where temperatures increase rapidly just above the surface.

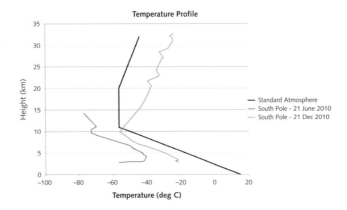

locations in the Arctic provides one explanation as to why the Antarctic is so much colder than the Arctic. The topography of the Antarctic continent, which is characterised by a high plateau over much of the interior, with steeply sloping terrain near the edge of the continent, is also important in shaping the winds that are a defining characteristic of the Antarctic climate.

The extremely cold temperatures are not the only unique aspect of temperature in the Antarctic. Temperature normally decreases with increasing altitude in the troposphere and this decrease in temperature is due to the fact that the surface of the Earth is strongly warmed by absorbed sunlight. Since only a limited amount of sunlight reaches the surface of the Antarctic and much of it, more than 80% in some locations, is reflected back out to space, there is very little warming of the Antarctic surface. In addition the surface is emitting longwave radiation, which leads to a net loss of energy from the ground and cooling of the surface. In the Antarctic the ground then serves to cool the air in contact with it rather than to warm the air as occurs in more temperate locations.

The situation described above occurs during the Antarctic summer months. In the winter, no solar radiation reaches the ground in the Antarctic and the cooling of the surface due to longwave radiation emission is even greater than in summer. As a result the air in contact with the cold surface also becomes very cold, and a temperature inversion develops. As one moves up through the Antarctic atmosphere, away from the cold ground, temperatures increase. This temperature inversion is found in the lowest several hundred metres of the atmosphere, and is a characteristic of the Antarctic climate that persists year-round. The inversion is strongest, that is the temperature increases most rapidly with height, during the winter when the ground is losing the most energy and is weaker in the summer when there is some solar radiation available to offset the longwave radiation loss. In the winter it is possible for the temperature to increase 25°C or more from the ground to an altitude of several hundred metres. Much of this temperature increase occurs close to the ground and it is possible that if you were

Box 4.2 Temperature inversions and Fata Morgana

The rapid increase of temperature just above the surface of the Antarctic results in a mirage known as a Fata Morgana. Many people are familiar with mirages that occur on hot, sunny days when a distant road appears to disappear in a distant lake. This mirage occurs when the ground is much warmer than the overlying air and results in light from the blue sky being bent as it passes through this temperature gradient. The result of this bending is that to a distant observer the blue light from the sky appears to come from below, and is seen as a shimmering water surface. The strong temperature inversion present over the Antarctic, especially during the winter, results in a similar mirage, except light is bent such that light reflected from the ground appears to be coming from the sky. The end result is that objects in the distance appear to be stretched vertically. Early polar explorers were tormented by these mirages as distant icebergs or other features of the distant landscape appeared to be looming ice palaces or cities. The appearance of these floating cities gave rise to the name Fata Morgana, which is Italian and relates to the fairy Morgan in the tales of King Arthur. In these tales the fairy was able to create castles in the air similar to the ice cities seen by early polar explorers.

Figure 4.11 Photograph of a Fata Morgana, a complex mirage – the apparent cliff at the base of the mountains – that appears lying just above the horizon and which develops in areas of strong temperature inversions. (Credit: John Cassano)

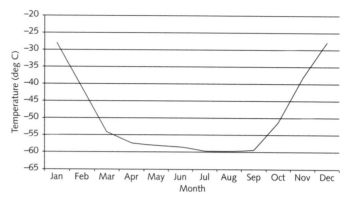

Figure 4.12 Annual cycle of monthly mean temperature at the South Pole. The coreless winter of uniformly cold temperatures from April to September is clearly shown in this plot.

standing on the high plateau of the Antarctic continent during the winter that the temperature could increase 5°C or more from your feet to the top of your head.

Another unique feature of the cold Antarctic winter temperatures is their duration. In mid-latitudes temperatures gradually decrease through the autumn with the coldest temperatures of the year confined to one or two mid-winter months. In the Antarctic, once the sun sets in March, temperatures quickly drop in early winter and remain nearly constant for 4 to 5 months of the Antarctic winter, at extremely cold temperatures, only to rise once the sun returns in the spring. Early Antarctic meteorologists called this unique seasonal change in temperature with a long period of nearly constant very cold temperature the 'coreless winter'.

Wind: why is Antarctica so windy?

Everyone knows that warm air rises. This is true in the atmosphere and is how thunderstorms form. Analogous to this is the concept that cold air sinks. This rising or sinking motion has to do with the density of the air. As air cools it becomes more dense and as it warms it becomes less dense. In a fluid, where fluid refers to both liquids and gases, less dense objects rise and denser objects sink. This can be seen with a glass of water. An ice cube placed in the glass of water will float because ice is less dense than liquid water. Similarly if a pebble is placed in the glass of water it will sink since the pebble is denser than the liquid water. In the atmosphere horizontal temperature contrasts lead to density contrasts which result in less dense warm air rising and denser cold air sinking.

Consider a location over the sloping edge of the Antarctic ice sheet. Air in contact with the ice sheet will cool by contact with the radiatively cooling ice surface. As one moves horizontally away from the ice surface the air becomes warmer. This horizontal temperature contrast creates a density contrast, with dense air adjacent to the ice sheet and less dense air further from the ice sheet. This dense air will sink, accelerating down the surface of the ice sheet to produce a wind known as a katabatic wind.

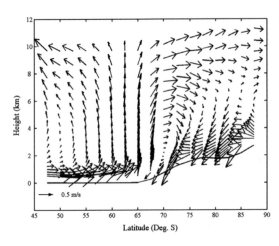

Figure 4.13 Schematic of the north–south and vertical winds over the Antarctic and mid-latitudes of the southern hemisphere. The Antarctic katabatic winds are seen as winds blowing from 90°S to 65°S in the lowest part of the atmosphere. Where these winds encounter the southward flowing mid-latitude winds strong rising motion occurs, which creates cloudy conditions near the Antarctic coast.

Katabatic winds are a defining characteristic of the Antarctic climate and are common over much of the Antarctic due to the coincidence of very cold air near the surface and the presence of sloping terrain over much of the continent. These katabatic winds initially flow straight downhill, much as water flows straight downhill, but eventually this wind is turned to the left due to the rotation of the planet. The result is air that spirals down and to the left from the high interior of the continent. Just as with flowing water the katabatic winds are channelled by the underlying ice topography resulting in them converging into 'valleys' in the ice sheet. The speed of these winds increases with decreasing temperature and with increasing terrain slope, but will also increase at locations where air from a large region is channelled through a small area. One of these locations is at Cape Denison, where Sir Douglas Mawson's Australasian Antarctic Expedition spent two years from 1912 through 1913. As the quote at the start of this chapter indicates, the winds at this location were unrelenting. Peak winds in excess of $50\,\mathrm{m\,s}^{-1}$ were not uncommon and the highest monthly mean wind speed observed by Mawson's party was $35\,\mathrm{m\,s}^{-1}$.

Adding to the misery of the strong winds experienced by Mawson's party is the fact that strong winds over a snow surface can result in dense blowing and drifting snow and blizzard conditions. A blizzard occurs when strong winds result in blowing snow that reduces horizontal visibility, and can occur even if no new snow is falling. The frequent blizzard conditions which Mawson's party experienced led Mawson to refer to Cape Denison as the 'Home of the Blizzard'.

The combination of steep topography and cold, dense air which engenders katabatic winds can also cause barrier winds to develop. A barrier wind occurs when cold, dense air blows towards a barrier, such as a high mountain range. Easterly winds persist around the edge of the Antarctic continent. These winds eventually encounter the Antarctic Peninsula, a steep mountainous barrier. The cold easterly winds are slowed and eventually blocked by this terrain, increasing the air pressure

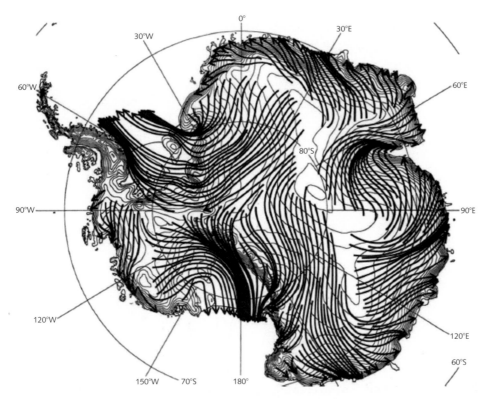

Figure 4.14 Map of Antarctic surface winds. These winds blow from the highest areas of the continent towards the coast and are channelled into regions of strong winds at several areas around the coast. (Credit: T. R. Parish and D. H.Bromwich)

Box 4.3 **Drifting snow and Antarctic research stations**

Blowing and drifting snow is still a concern for present Antarctic scientists. The presence of any structure in an area of blowing snow will result in a disruption of the wind and lead to areas of snow accumulation. This snow accumulation can bury buildings, making them uninhabitable. This problem occurs at many Antarctic research stations. The United States base at the South Pole had occupied a dome-like building for many years, but eventually this dome was slowly buried by accumulating snow and has now been replaced by new buildings. The British station Halley has had a similar problem of being buried by the accumulating snow year after year. Halley has been rebuilt six times, with each station built on the snow surface lasting less than 10 years before being buried and becoming uninhabitable. The newest Halley Station is built of modules on skis that allow the buildings to be towed to new locations as snow accumulates around them.

Box 4.4 **Observing Antarctic weather and climate: automatic weather stations**

Figure 4.15 Photograph of the Sabrina automatic weather station on the Ross Ice Shelf. (Credit: Shelley Knuth, Cooperative Institute for Research in Environmental Sciences, University of Colorado)

There are very few year-round stations on the continent so that vast areas of the continent are unoccupied with nobody to collect weather data. Starting in the early 1980s automatic weather stations measuring temperature, pressure, wind speed and direction, and humidity have been installed in many remote locations. The weather stations can operate for years without being visited, although sometimes the harsh Antarctic climate does damage or destroy these weather stations. The data collected

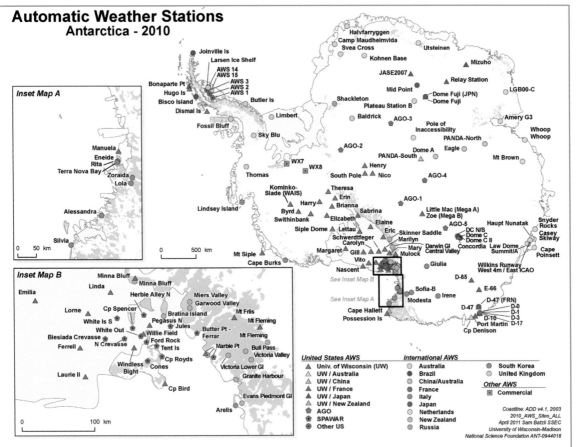

Figure 4.16 Map of current automatic weather station locations. (Credit: Matthew Lazzara, Antarctic Meteorological Research Center, Space Science and Engineering Center, University of Wisconsin–Madison)

are transmitted to polar orbiting satellites, and rebroadcast around the world in real-time. This data is used to make forecasts of Antarctic and Southern Ocean weather as well as being crucial for studying trends in the climate of the Antarctic and Southern Ocean. Currently over 100 automatic weather stations are maintained by several nations. While these weather observations are obviously important for forecasting the weather in and around Antarctica, they are also critical for weather forecasting across the planet. The computer models that meteorologists use to forecast the weather must have information about the current state of the atmosphere before they can produce a forecast of future weather conditions. Many of these computer models cover the entire planet, and the weather data collected in the Antarctic is used in these models.

near the base of the mountains. This increase in pressure results in air starting to flow away from the barrier, opposite to its original direction, from higher pressure towards lower pressure. Under the influence of the planet's rotation the westerly wind turns to the left, eventually turning parallel to the topography, and becoming a southerly wind. These terrain parallel winds are referred to as barrier winds. Southerly barrier winds are common on the eastern side of the Antarctic Peninsula and on the east side of the Transantarctic Mountains, which form the western boundary of the Ross Ice Shelf.

Precipitation: why is Antarctica so dry?

On average Antarctica is the driest continent, with average annual precipitation of 13 cm, and can be considered a desert. Why is Antarctica so dry? As discussed earlier, the Antarctic continent is located in an area of sinking motion, associated with the polar cell. Sinking motion in the atmosphere tends to cause clouds to dissipate, and suppresses the development of precipitation. This, in large part, is why Antarctica gets so little precipitation. In addition the atmosphere tends to contain less water vapour as temperature decreases and as the coldest continent the Antarctic atmosphere also contains very little water vapour.

The fact that Antarctica can be considered a desert may seem surprising given that 99% of the continent is permanently covered by ice. How could thousands of metres of ice come to exist on the Antarctic continent when just over 1 cm of precipitation falls per year in the interior of the ice sheet? The answer to this riddle is that what little snow does fall over the interior of the continent does not melt, and thus can accumulate year after year, century after century, and millennia after millennia. Over these long periods the small annual snowfall can accumulate to great depths, resulting in an ice sheet that is more than 4 km thick at some locations. In fact, ice retrieved from ice cores at Concordia Station on the East Antarctic plateau has been found to be over 720 000 years old.

If Antarctica is the driest continent, in terms of annual precipitation amount, is the continent dry or humid when one considers the amount of water vapour in the Antarctic atmosphere? The answer to this question is that it is both very dry and also quite humid, at the same time and place. How can this be? When meteorologists discuss humidity they can refer to either absolute or relative measures of humidity. An absolute measure of humidity is an indication of the actual amount of water vapour present in the atmosphere. From this perspective Antarctica is quite dry. As air temperature decreases the air is able to contain less water vapour. At the very cold temperatures found in the Antarctic atmosphere during the winter the air can contain less than 1% of the water vapour that could be contained in air on a warm summer day in the mid-latitudes.

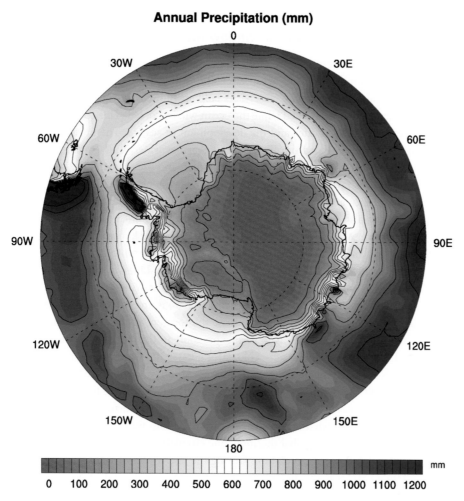

Annual Precipitation (mm)

Figure 4.17 Map of Antarctic precipitation distribution. (Credit: Matthew Higgins, Cooperative Institute for Research in Environmental Sciences, University of Colorado)

The relative measure of humidity used by meteorologists, known simply as relative humidity, indicates how much water vapour is in the atmosphere relative to the maximum amount of water vapour the atmosphere could contain at that temperature. The relative humidity can be viewed as being similar to the fuel gauge in a car. When the relative humidity is 0% the atmosphere is completely dry, and corresponds to the car's fuel gauge indicating that the fuel tank is empty. When the relative humidity is 100% the atmosphere is said to be saturated and it contains the maximum amount of water vapour possible. In this situation fog or clouds will form. This situation corresponds to the fuel gauge indicating that the fuel tank is full. The relative humidity can change if either the amount of water vapour in the

Box 4.5 Antarctic polynyas: a window into the ocean

We have seen that the temperature gradient between the poles and the tropics drives the global atmosphere and ocean circulations, but these two parts of the climate system also interact quite closely. As cold air blows over the ocean the air will be warmed and moistened while the ocean will be cooled. This cooling can lead to freezing of the seawater and the development of sea ice. The presence of sea ice insulates the cold air from the relatively warmer water, thus limiting the exchange of heat and moisture between the ocean and the atmosphere. While sea ice formation is favoured by cold air flowing over the ocean, which is a common occurrence around the margins of the Antarctic continent, very strong winds can also push this sea ice away from the Antarctic coast leading to areas of open water adjacent to the continent. These areas are known as polynyas, and form in similar locations year after year, where winds tend to be strongest. One of these locations is Terra Nova Bay in the Ross Sea. Like Cape Denison, Terra Nova Bay is subject to exceptionally strong katabatic winds. Members of Scott's Antarctic expeditions that spent the winter on an island in Terra Nova Bay named the island Inexpressible Island, because the torment of the strong winds day after day was so unbearable as to be inexpressible to those that had not experienced it at first hand. The very cold, strong katabatic winds which blow across Terra Nova Bay remove significant amounts of heat and moisture from the ocean. As a result the ocean is cooled, and its salinity increases due to the evaporation of water. This cold, salty water is denser than the surrounding ocean water and sinks to form what is known as Antarctic bottom water. This Antarctic bottom water is part of the global thermohaline circulation, which is responsible for driving the global ocean circulation. During the late winter of 2009, scientists from the United States made measurements of the atmosphere and ocean over the Terra Nova Bay polynya, to understand better the processes acting in this unique environment. Since flight operations and fieldwork are limited to the more mild Antarctic summer this research was conducted using a remotely controlled aircraft, known more formally as an unmanned aerial vehicle (UAV). The UAV carried instruments to measure the atmosphere and to observe the surface of the polynya. Measurements from the UAV showed white-capped waves being driven by hurricane force winds and heat and moisture transfer from the ocean that were as large as the heating that could be expected over a mid-latitude location in summer. These observations will allow scientists to more accurately represent the heat and moisture exchange between the ocean and the atmosphere in climate models, and to better characterise the properties of the Antarctic bottom water that forms here. British scientists have used a similar UAV to make measurements over the Weddell Sea during the Antarctic summer (October to December) of 2007.

Figure 4.18 (right) Photograph of an Aerosonde unmanned aerial vehicle launch. (Credit: John Cassano)

Figure 4.19 (below) Aerial photo of Terra Nova Bay polynya from UAVs. The blue regions are open water, with white-capped waves. The bands of white are areas of loosely consolidated sea ice. The photograph shows a section 200 m wide. (Credit: John Cassano)

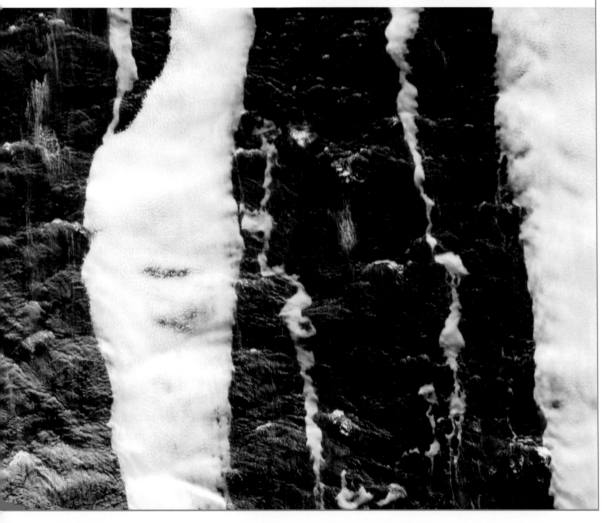

Figure 4.20 Meteorologist Bob Bedner releases a weather balloon at Amundsen–Scott South Pole Station. These balloon launches provide a profile of the atmosphere and aid with weather forecasts. (Credit: Phillip Marzette, NSF)

atmosphere changes or if the temperature of the air changes. Since this maximum amount of water vapour decreases rapidly with decreasing temperature, at very cold temperatures the atmosphere only needs to contain a very small amount of water vapour in order to be saturated. As a result the Antarctic atmosphere can be dry in an absolute sense, that is it can contain very little water vapour, but because the temperature is so cold it can also have a very high relative humidity. The relative humidity over the Antarctic continent is often 90% or greater.

Weather forecasting in the Antarctic

Modern weather forecasting, whether for the Antarctic or anywhere else on our planet, relies on a combination of current weather observations, provided by surface weather stations, weather balloons (radiosondes), and satellite observations, predictions from computer models of the atmosphere, and human forecasters. Uncertainties exist with all of these elements in the Antarctic, increasing the difficulty of producing accurate weather forecasts.

There are approximately 100 surface weather observation points in the Antarctic while there is an order of magnitude more weather observations over the United States or Europe, both of which cover a smaller area than Antarctica. Since the atmosphere is three-dimensional, weather forecasters need to consider not only the surface weather but weather data through the depth of the troposphere. A key method of obtaining this data is from radiosondes. Radiosondes are routinely launched twice per day and make measurements of temperature, pressure, humidity, wind speed and wind direction. There are fewer than 20 radiosonde sites in the

Figure 4.21 Map of Antarctic radiosonde locations. (Credit: Matthew Lazzara. Antarctic Meteorological Research Center, Space Science and Engineering Center, University of Wisconsin–Madison)

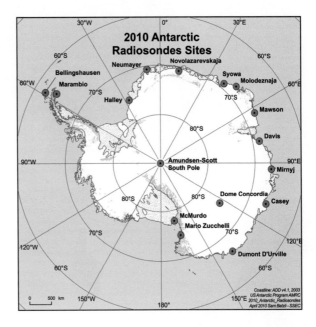

Antarctic, which is also almost an order of magnitude less than the number of radiosonde observations in the United States or Europe.

Satellites provide critical data for weather forecasting, including obvious information such as the location of cloudy areas and the position of storms and fronts, but less obviously information of temperature, humidity and winds. One problem with satellite data over the Antarctic is that it is often difficult to distinguish between cloudy and cloud-free areas, since both the clouds and the underlying ice surface are both highly reflective. This difficulty in identifying clouds also makes accurate retrieval of temperature and humidity values from satellites difficult. Finally, due to the orientation of satellite orbits polar orbiting satellites only pass over the polar regions several times per day unlike tropical and mid-latitude locations where geostationary satellites make continuous observations. Taken together this situation results in a substantially reduced number of weather observations in the Antarctic compared to tropical or mid-latitude locations, which has serious implications for Antarctic weather forecasting.

Computer models, known as numerical weather prediction models, are another key element of weather forecasting. These models are based on the mathematical equations that describe the behaviour of the atmosphere. These equations are based on the basic physics that controls the atmosphere as well as a number of assumptions. These equations can only be evaluated at a limited number of points (often hundreds of thousands of points) due to limitations on available computer power. This limited number of points means that the forecasts from these numerical weather prediction models can only represent atmospheric features at scales larger

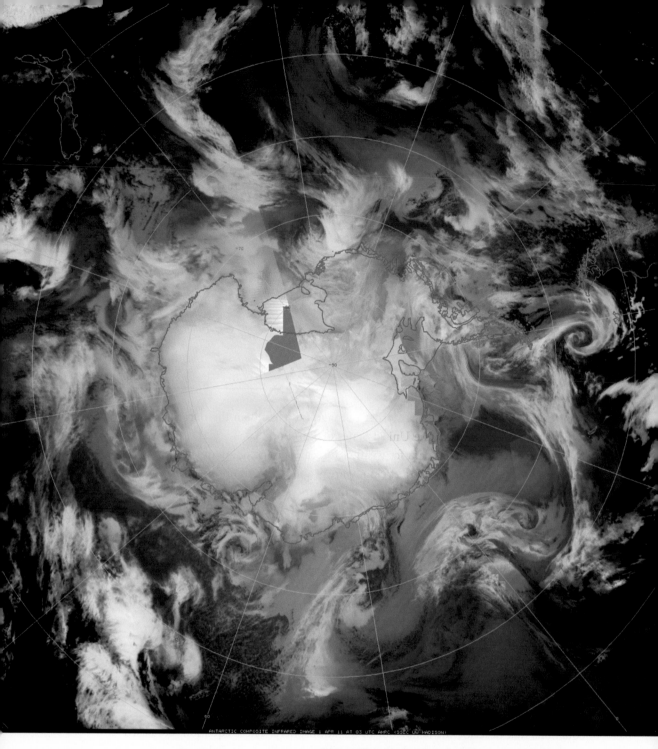

ANTARCTIC COMPOSITE INFRARED IMAGE 1 APR 11 AT 03 UTC AMRC (SSEC UW-MADISON)

Figure 4.22 Antarctic infrared satellite image (composite). This image shows the temperature of the surface (where skies are clear) or cloud tops. Cold regions are shown by light shading and warm regions are shown by dark shading. (Credit: Matthew Lazzara. Antarctic Meteorological Research Center, Space Science and Engineering Center, University of Wisconsin–Madison)

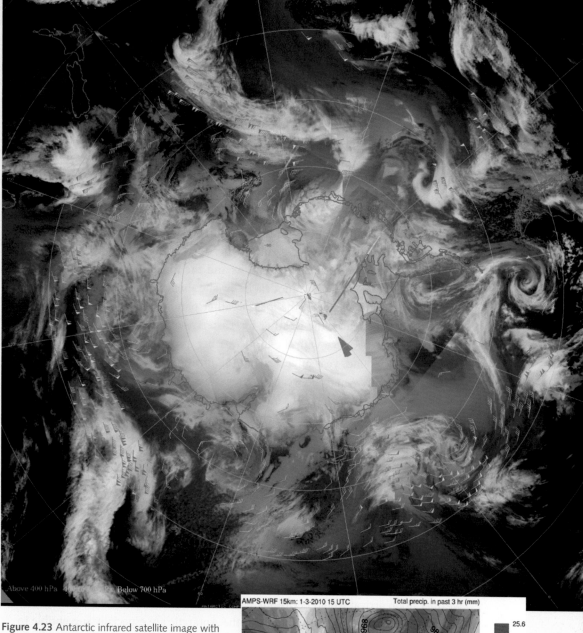

Figure 4.23 Antarctic infrared satellite image with derived winds. Colours on the wind barbs indicate winds at different heights in the atmosphere. (Credit: Matthew Lazzara. Antarctic Meteorological Research Center, Space Science and Engineering Center, University of Wisconsin–Madison)

Figure 4.24 Forecast map from AMPS showing sea level pressure (blue lines) and precipitation (green shading). (Credit: John Cassano)

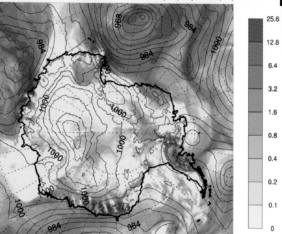

than the spacing of the points in the model, limiting the accuracy of the weather forecasts. Another necessary ingredient for successful weather prediction is an accurate representation of the current atmospheric state and with the limited atmospheric observations available in the Antarctic, this is often a problem. A final issue is that since some of the Antarctic atmospheric processes represented by the equations differ from those in mid-latitude locations these models are not ideally suited for Antarctic weather forecasting. Over the last decade modifications to mid-latitude numerical weather prediction models have considerably improved numerical weather prediction in the Antarctic.

The final ingredient for successful weather forecasting is the human forecaster. Using all of the atmospheric observations and the forecasts from one or more numerical weather prediction models he or she then modifies it in the light of experience and knowledge of the local weather patterns. Since no one lives in Antarctica permanently, the weather forecasters that work there often do not have as much experience with the local weather as a forecaster in the mid-latitudes who lives year-round, year after year, in the location where they are forecasting. Fortunately, many Antarctic weather forecasters return to the Antarctic for many field seasons, and thus do gain knowledge of the local weather and experience in forecasting for this challenging environment.

Through the addition of new automatic weather station locations around the continent, improved satellite coverage and data processing, and advances in numerical weather models the accuracy of Antarctic forecasts has improved dramatically over the last 20 years. In the early 1990s forecasting beyond 8 hours was a problem, whereas now accurate forecasts for 3 to 5 days are commonplace. This change has had significant implications for those travelling to and from Antarctica for work or for tourism, since flights to Antarctic often take 5 to 8 hours. In the early 1990s pilots flying to Antarctic would receive weather forecasts for their arrival that were just at the edge of what weather forecasters considered reliable. As a result the weather upon arrival would occasionally differ, sometimes quite dramatically, from what was forecast. Aircraft operations in the Antarctic are especially susceptible to strong crosswinds and reduced visibility due to low clouds, fog or blowing snow. If conditions at the Antarctic landing site are too bad the incoming aircraft is unable to land and must return to its departure airport, a situation known as a boomerang flight.

Climate change in the Antarctic

Human induced, or anthropogenic, climate change has already had a significant impact on the Antarctic through the development of the ozone hole. It is also likely that changes in the Earth's climate due to the increase of greenhouse gases in the atmosphere will alter the Antarctic climate over the next decades to centuries.

Box 4.6 **Weather forecasting for a mid-winter rescue flight to the South Pole**

In addition to crosswinds and reduced visibility, the extremely cold temperatures at the South Pole also present a significant aviation hazard. At very cold temperatures the hydraulic fluid used in aircraft to operate rudders and wing flaps begins to congeal and cannot be used, making landing or taking-off impossible. As a result of the cold temperatures, the normal flight season at the South Pole is quite restricted (October to February). When the last flight leaves in February no further flights will arrive for 9 months. During the winter of 2001, the doctor at South Pole station, Dr Ronald Shemenski, began to suffer from pancreatitis and gallstones. Medical treatment for this was not possible on the station, and he needed to be evacuated during the winter.

The United States Antarctic Program contracted with Kenn Borek Air Ltd. of Canada to carry out the rescue flights using ski-equipped Twin Otter aircraft. Two Twin Otter aircraft departed Canada in April of 2001 and flew south to Punta Arenas, Chile and then on to the British station of Rothera on the Antarctic Peninsula. The flight from Rothera to the South Pole is a 10-hour flight in a Twin Otter, and the aircraft waited at Rothera for a forecast of suitable weather for a round trip. Both United States and British weather forecasts relied on weather observations and a suite of numerical weather prediction models to make their forecast for this unprecedented mid-winter flight to the South Pole. One of the Twin Otter aircraft arrived at the South Pole on 24 April, with the second Twin Otter aircraft staying at Rothera to provide support in case of problems. The Twin Otter returned to Rothera, with Dr Shemenski on 26 April, approximately 34 hours after first leaving Rothera. This mid-winter rescue flight would not have been possible without an accurate weather forecast that was able to predict a lull in the winds, and blowing snow at the South Pole that lasted only slightly longer than the length of the mission.

The Antarctic ozone hole

In 1985 British scientists announced the discovery of reduced amounts of ozone over the Antarctic continent during the spring months. Surface-based observations of ozone conducted at Halley station indicated that ozone amounts had been decreasing since 1977, but there was concern that satellite measurements of ozone made during the same time period did not indicate a similar decrease. Upon closer inspection of the satellite data it was found that the algorithm used to process the satellite data was rejecting the low ozone values observed by the satellite as the values were outside of the 'normal' range of values. It was in this way, through the use of multiple observing systems, that the world first became aware of the Antarctic ozone hole.

How does the ozone hole develop and why is it only observed over the Antarctic?

Before answering these questions we should first explain what is meant by the term 'ozone hole'. The ozone hole represents a near total depletion of ozone within a portion of the stratosphere and develops each year during the Antarctic spring months of September and October.

In order to understand why the ozone hole forms over the Antarctic it is useful to first understand how ozone can be destroyed in the atmosphere. Ozone molecules break down naturally when they absorb ultraviolet radiation, but they can also be destroyed by chemical reaction with atoms of chlorine. It has been estimated that one atom of chlorine is capable of destroying 100 000 ozone molecules. One atmospheric source of chlorine is chlorofluorocarbons (CFCs), previously in wide use as a propellant for aerosol spray cans and in refrigerants up to the 1990s. CFCs have a lifetime in the atmosphere of approximately 100 years and though released in the lower portion of the atmosphere will slowly mix upward, eventually reaching the stratosphere. In the presence of ultraviolet radiation CFCs are broken down, freeing chlorine atoms, which can then destroy ozone. The process described above occurs in the stratosphere all around the world. However, a unique combination of events takes place over the Antarctic stratosphere to produce the rapid spring depletion of ozone.

The stratosphere is normally heated by the absorption of ultraviolet radiation, but during the Antarctic winter no solar radiation, and thus no ultraviolet radiation, is present and the stratosphere becomes quite cold. In addition, the lower portion of the atmosphere is also cold at this time of the year due to longwave radiation losses at the surface. These very low temperatures throughout the atmospheric column cause the formation of the polar vortex, a band of very strong westerly winds that encircle the Antarctic and limit the amount of mixing of air between Antarctica and lower latitudes. This lack of mixing drives temperatures down further so that the temperature in the Antarctic stratosphere can drop to $-90°C$. At temperatures below $-85°C$ polar stratospheric clouds, also known as nacreous clouds, can develop. Unlike 'normal' tropospheric clouds, which are composed of liquid water and ice, polar stratospheric clouds are composed of ice and nitrogen compounds. Chemical reactions on the particles which make up these clouds lead to the breakdown of CFCs and the formation of molecular chlorine. Unlike atomic chlorine, molecular chlorine does not destroy ozone. The presence of polar stratospheric clouds during the Antarctic winter leads to a destruction of CFCs and an increase in the amount of molecular chlorine in the Antarctic stratosphere.

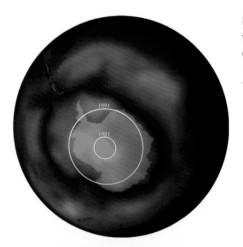

Figure 4.25 Map of ozone concentration showing low ozone values (red) over the Antarctic in September 2011. The dated circles show the earlier extent of the ozone hole in 1981 and 1991. (Credit: NOAA)

Figure 4.26 Photo of polar stratospheric clouds (also known as nacreous clouds) over McMurdo Sound. (Credit: John Cassano)

At the end of winter, when sunlight returns, the molecular chlorine is broken down by the energy provided in ultraviolet radiation to form atomic chlorine. This atomic chlorine then rapidly depletes ozone leading to the formation of the ozone hole. The minimum amount of ozone observed over the Antarctic had decreased by a factor of two from the early 1980s until the early 2000s, but there is hope that this trend will reverse.

Upon realising that CFCs were responsible for depleting stratospheric ozone an international agreement, known as the Montreal Protocol, was signed in 1987 to

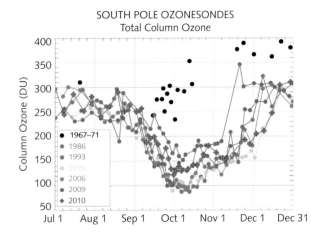

SOUTH POLE OZONESONDES
Total Column Ozone

Figure 4.27 Time series of minimum ozone concentration over Antarctica (Credit: NOAA)

limit and eventually phase out the production of CFCs. This protocol has begun to show signs of taking effect, with a decrease in CFCs in the stratosphere since 1997. Given the long lifetime of CFCs in the atmosphere a reversal of the ozone hole is not expected immediately, but may begin to become evident by as early as 2015. It is possible that the spring ozone hole will no longer develop by the end of the twenty-first century. The fact that the world's nations were able to act to reverse a global environmental problem like the ozone hole provides hope that a solution for global warming can be achieved as well.

Global climate modelling and global warming

Observations of carbon dioxide at Mauna Loa observatory in Hawaii and at the South Pole in Antarctic, since 1957, have shown a 20% increase in the amount of carbon dioxide in the atmosphere. Scientists have become increasingly concerned over the last two decades about the possible impacts of increasing greenhouse gases (carbon dioxide, methane and others) on the climate system. The concern is that an increase in greenhouse gases in the atmosphere, primarily due to the burning of fossil fuels, will alter the radiation balance of the atmosphere since these gases are effective in absorbing longwave radiation. In doing this these gases trap longwave radiation emitted from the surface of the Earth, resulting in a warming of the lower atmosphere and an increase in surface temperature. Further, there is reason to believe that these changes in temperature will also be accompanied by changes in precipitation and atmospheric circulation patterns. These changes will also impact other aspects of the climate system, such as monsoons, hurricanes, sea ice and ice sheets.

The United Nations Intergovernmental Panel on Climate Change (IPCC) has been tasked with evaluating observed climate change, assessing what role human

Observations

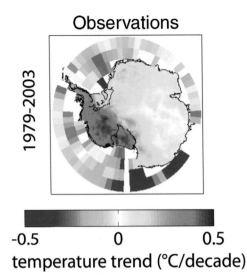

1979–2003

-0.5   0   0.5

temperature trend (°C/decade)

Figure 4.28 Map of observed twentieth-century warming over Antarctica. (Credit: Steig E. J., D. P. Schneider, S. D. Rutherford *et al.* (2009). Warming of the Antarctic ice-sheet surface since the 1957 International Geophysical Year. *Nature*, **457**, 459–464.)

activities have had in any observed change and estimating what future changes in the climate system may result from human activity. The most recent IPCC report, the Fourth Assessment Report, was released in 2007. The authors of this report were awarded the Nobel Prize in 2007 for their contribution to this globally important effort to understand climate change. Conclusions from this report indicate that from 1906 to 2005 the globally averaged surface temperature has warmed 0.74 ± 0.18°C, with the rate of warming larger during the latter 50 years of this period. Further, they determined that it is very likely that the warming over the past 50 years is a result of increasing greenhouse gas concentrations in the atmosphere.

How do atmospheric scientists determine if observed changes in global temperature are due to increasing concentrations of greenhouse gases in the atmosphere and how do they make projections of future climate change, given that the climate system is extremely complex? In part scientists have turned to the use of global climate system models to address these questions. These models are computer programs that represent processes such as the transfer of longwave and shortwave radiation through the atmosphere and the response of these radiative transfers to changes in greenhouse gases. These models also represent the atmospheric circulation, and other atmospheric physical processes, similar to numerical weather prediction models. Finally, these models include details on other parts of the climate system such as the oceans, sea ice and land surfaces. These models have been shown to reasonably reproduce the current state of the climate system. Given this fact climate scientists have confidence in the fidelity of these models, and can then use these models to explore 'what-if' scenarios.

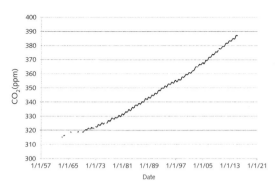

Figure 4.29 Time series of South Pole carbon dioxide concentration. (Credit: Ralph Keeling, Scripps Institution of Oceanography)

One type of what-if scenario that has been explored using global climate system models is how the global temperature during the twentieth century would have varied without the observed increase in greenhouse gases. Results from these experiments clearly demonstrate that the observed warming during the twentieth century can only be reproduced in the climate system model simulations when the observed increases in greenhouse gases are included in the models. Without these increases in greenhouse gases little change in global temperatures is simulated by the models, inconsistent with the observed changes in global temperatures.

A second what-if question is: what is the impact of changing the amount of greenhouse gases in the atmosphere over the next century? Evaluation of global climate system model simulations, and other types of models, forced with a range of possible changes in greenhouse gas concentrations, indicate that the global average temperature is likely to warm between 1.1 and 6.4°C by the year 2100. Based on the IPCC Fourth Assessment Report, it is likely that surface temperatures and precipitation in the Antarctic will increase during the twenty-first century.

In the Antarctic reliable weather observations extend back only to the middle of the twentieth century, unlike mid-latitude locations in the northern hemisphere where reliable weather observations can extend back several centuries. Based on a combination of Antarctic weather observations and satellite data surface temperatures over the continent have warmed by 0.12°C per decade from 1957 to 2006. This warming is greatest in West Antarctica, which has warmed 0.18°C per decade from 1957 to 2006. Precipitation data over the Antarctic is even more limited than temperature data, and no trends in Antarctic precipitation have been observed during the second half of the twentieth century.

Global climate system model projections of climate change in the Antarctic indicate warming of between 1.4 and 5°C during the twenty-first century. This warming is expected to be relatively uniform across the continent, in contrast to the very localised warming observed during the twentieth century. Precipitation is also expected to increase over the Antarctic continent. The IPCC Fourth Assessment Report indicates that precipitation is likely to change between −2% and +35% over

Figure 4.30 Intergovernmental Panel on Climate Change (IPCC) figure of simulated twentieth century temperatures (a) with (yellow and red lines) and (b) without (blue lines) greenhouse gas forcing. The black lines in (a) and (b) show the observed global temperature.

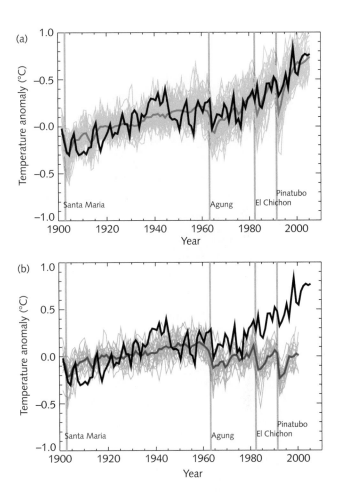

the Antarctic during the twenty-first century. This expected increase in Antarctic precipitation is important in that increased snowfall over the Antarctic continent will contribute to an increase in the amount of ice contained in the Antarctic ice sheet, since even with the projected warming Antarctic temperatures will still be too cold to allow melting over much of the ice sheet. This increase in ice stored in the Antarctic ice sheet will contribute to a reduction in global sea level, helping to offset part of the expected sea level rise from other sources during the twenty-first century.

In this chapter we have discussed the mechanisms responsible for creating the unique weather found in the Antarctic. The weather in Antarctica is unlike the weather anywhere else on our planet, with conditions ranging from the coldest temperatures recorded at the surface of Earth to locations that experience hurricane force winds for weeks on end. We have also seen that the weather and climate of Antarctica are critical to driving the weather and climate across our planet. At the most basic level, the contrast between the cold polar regions and the warm tropics

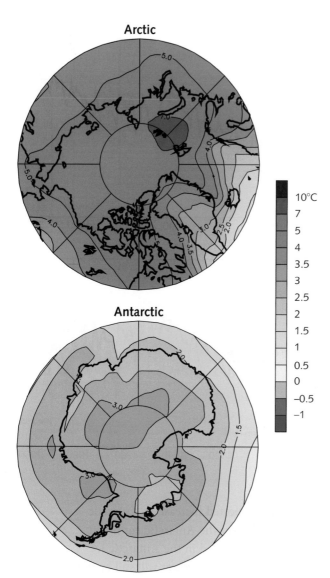

Figure 4.31 Intergovernmental Panel on Climate Change (IPCC) figure of projected temperature change at end of twenty-first century.

ultimately drives all of the atmospheric and oceanic circulations on Earth, and gives rise to our global climate as we know it. It would seem that projected changes in Antarctic climate during the twenty-first century, particularly expected increases in snowfall over the Antarctic ice sheets, will have an impact on changes in global sea level that impact all of us. Continued observation of the Antarctic atmosphere and research into the linkages between Antarctic weather and climate and that of lower latitudes and the interactions between the atmosphere, ocean, sea ice and ice sheets will allow scientists to better refine these estimates of climate change.

5 | Stormy and icy seas

EBERHARD FAHRBACH

> As we laboured south the storm continued with increasing violence, for now we were entering the 'Roaring Forties'. Our vessel staggered to the crest of mountainous seas and plunged heavily down long troughs. **Frank Hurley, 1925**

The geographical setting

The cold and stormy Southern Ocean surrounds the Antarctic continent and connects the southernmost parts of three major ocean basins – the Atlantic, the Indian and the Pacific Ocean. This stretch of water, although far from most centres of population, has a crucial effect on the global climate system through its role in the circulation of the world's oceans, the uptake of carbon dioxide from the atmosphere and, through its formation of annual sea ice, an influence on climate elements of importance to marine productivity.

Formally, its northern limit is now taken as 60°S, corresponding roughly to the southern boundary of the Antarctic Circumpolar Current and to the northern limit of sea ice in winter. Covering an area of 20.3 million km^2, its mean depth is 4500 m (with its greatest depth being 7235 m). However, to appreciate its role in the climate system, the complete Antarctic Circumpolar Current has to be included reaching as far north as 40°S. Then, the Southern Ocean covers 70 million km^2 which corresponds to 20% of the world's oceans. Where the Drake Passage separates Cape Horn on the South American side from the South Shetland Islands on the Antarctic side it is only 800 km wide, whereas the distance from South Africa (Cape Agulhas) to the Antarctic continent (Dronning Maud Land coast) is 3900 km. To the south the Southern Ocean is limited by the Antarctic coastline of 17 968 km which is largely covered by ice shelves.

Antarctica: Global Science from a Frozen Continent, ed. David W. H. Walton. Published by Cambridge University Press. © Cambridge University Press 2013.

ANTARCTICA AND SOUTHERN OCEAN

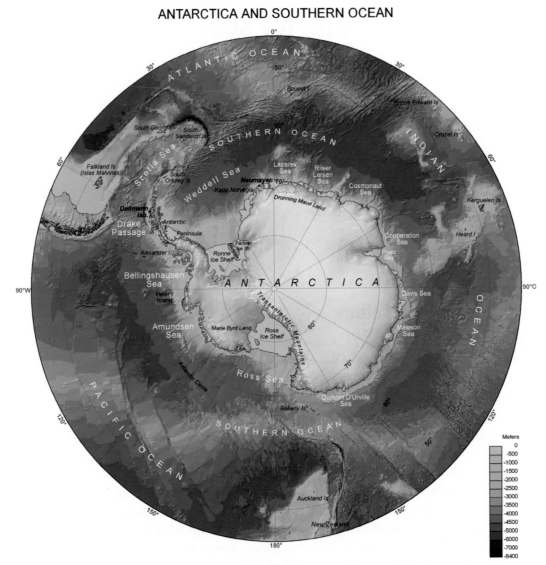

Cartography: D. Graffe, Alfred Wegener Institute. Modified from: IOC, IHO and BODC (2003), Centenary Edition of the GEBCO Digital Atlas, published on CD-ROM on behalf of the Intergovernmental Oceanographic Commission and the International Hydrographic Organization as part of the General Bathymetric Chart of the Oceans, British Oceanographic Data Centre Liverpool, UK.

Figure 5.1 Topographic map of the Southern Ocean displaying the deep basins and ridges which structure the ocean in hydrographic regimes. (Credit: H.-W. Schenke, AWI)

This enormous area is divided underwater by mid-ocean ridges into a series of basins with quite distinct and different water masses.

The Weddell Abyssal Plain is located east of the Antarctic Peninsula, being limited to the north by the South Scotia and the North Weddell Ridges whilst to the east of it, the

Enderby Abyssal Plain is bounded to the north by the Southwest–Indian Ridge. Further to the east, beyond the Kerguelen Plateau, extends the Australian–Antarctic Basin which is limited in the north by the Southeast Indian and the Indian–Antarctic Ridges. East of the Macquarie Ridge, the Ross and Amundsen Abyssal Plains and the Bellingshausen Plain lead into the Drake Passage, limited to the north by the Pacific–Antarctic Ridge and the Eltanin Fracture Zone. The ridges have a significant influence on the location of major currents and fronts.

The Southern Ocean formed about 120 million years ago during the Gondwana breakup when the South American, the African and the part of the Indo-Australian plates started to move north away from the Antarctic plate. The Pacific plate separated from the Antarctic plate about 100 million years ago and the Australian part of the Indo-Australian plate about 60 million years ago. After the Tasmania–Antarctic Passage had opened about 34 million years ago and the Drake Passage about 31 million years ago a circumpolar deep water flow existed for the first time and changed dramatically the climate and the biology of the whole of the southern hemisphere. As a direct consequence of the formation of these circumpolar wind and current systems, Antarctica cooled and the continent-wide ice sheet began to grow.

Winds and weather

The Southern Ocean is a very windy place. Winds are generated by horizontal air pressure differences which are caused by temperature and humidity gradients between the polar and the subtropical latitudes. To the north of the Southern Ocean (Figure 5.2) is the subtropical high pressure belt, southward from which the air pressure at the sea surface decreases to the low pressure trough which is centred at about 60°S. The pressure difference between the subtropical highs and the low pressure belt, together with the action of the rotating Earth, give rise to a band of strong west winds, the West Wind Drift. The low pressure belt consists of a series of moving cyclonic systems which are continuously forming and decaying. South of the low pressure belt, pressure increases again towards the continent which results in easterly winds in the Antarctic coastal areas.

Thus the West Wind Drift drives the eastward flowing ocean current, the Antarctic Circumpolar

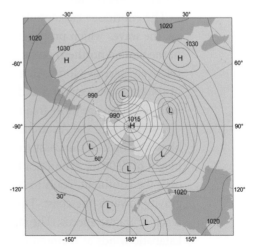

Figure 5.2 Atmospheric pressure distribution at sea level displaying the major high and low pressure systems, which determine the winds over the Southern Ocean. (Credit: Eberhard Fahrbach, AWI)

Figure 5.3 Waves in the Southern Ocean breaking over the ship's bows. (Credit: Peter Bucktrout, BAS)

Current, whilst the westward flowing Antarctic Coastal Current is related to the East Wind Drift.

The West Wind Drift, with its frequently very strong winds and high waves, has been a problem for sailors and explorers since the early days of exploration and navigation in the southern hemisphere. Between 45°S and 55°S where the strongest winds are found, gale force winds can occur as frequently as 20% of the time. The sailing conditions of the Southern Ocean are reflected by terms such as 'Roaring Forties', 'Furious Fifties' and 'Screaming Sixties'. Since the low pressure systems are of the same size (around 1000 km across) and the same intensity in both the southern and northern hemispheres, and since the wave height depends on the distance and the time over which the wind is blowing in the same direction, these wind-generated waves are actually of a similar height in the Southern Ocean and the North Atlantic or North Pacific. The largest wave heights normally to be expected in the West Wind Drift are 25 m but under particular conditions even larger waves may be formed. In certain areas, for example at the southern tips of South Africa and South America, superposition of waves originating from different wind systems can interact with ocean currents to generate extremely high waves, monster waves. South of Africa interference of waves from the Indian and the Atlantic Oceans might lead to wave heights of 35 m. Most likely such 'monster waves' occur more often than was believed in the past based on ship's observations only, but satellite detection of such events started only recently.

The enormous circumpolar extent of the Southern Ocean allows the swell, i.e. the waves which propagate out of the wind-driven area, to travel farther than in the limited northern oceans before reaching coasts. This holds in particular for the relatively narrow North Atlantic. Swell higher than 4 m is present for 30 to 70% of the time between 50°S and 60°S. Since the winds are weaker in summer, high seas and swell are most frequent in autumn, winter and spring.

The winds in the Antarctic coastal area are strongly influenced by the winds blowing down to the coast from the interior of the continent. With an average height of 2500 m, Antarctica is the highest of all continents. High pressure air sinks over the continent, cools to become denser and then flows steadily down-slope by gravity towards the coast. This down-slope motion, the katabatic winds, can reach dramatic speeds. The strongest winds ever measured in Antarctica were the katabatic winds at Dumont d'Urville with a speed of 327 km h^{-1}. The superposition of the East Wind Drift and katabatic winds generate offshore wind patterns that are of great importance to sea ice formation.

Water mass formation

Temperature and the amounts of dissolved salts (salinity) characterise the water masses in any ocean. Since they are not scattered randomly but occur in well-defined combinations this allows us to determine how each water mass gained its specific properties. Water types are defined by a pair of temperature and salinity measurements and show up as points in a temperature/salinity plane, the so-called T/S-diagram. Two water types mixing with each other are found on a line in the T/S-diagram.

In addition to temperature and salinity, water masses can be distinguished by their different content of dissolved substances, gases such as oxygen or carbon dioxide, or nutrients, trace metals or anthropogenic substances such as CFCs.

Temperature and salinity of the oceans vary in both space and time (Figures 5.4 and 5.5). They are determined by heat and fresh water exchanges at the ocean surface. The ocean gains heat by solar radiation or heat transfer from the warmer atmosphere and loses heat by radiation, contact with the colder atmosphere or by evaporation. Freshwater is gained from precipitation, river inflow and melting ice and is lost by evaporation and sea ice formation. Since the ocean gains heat in low latitudes and loses it in high latitudes, its temperature at the surface normally decreases from low to high latitudes. Since there is an excess of evaporation in the subtropics and of precipitation further north and south, there is generally a decrease of sea surface salinity from the subtropics to the north and to the south. A water mass originating in the subtropics would be warm and saline in comparison to one from the mid-latitudes which would be colder and fresher. Ocean waters of a given characteristic are transported by ocean currents out of the area where they formed, initially forming tongues of warm or cold water but later mixing and losing their individuality.

Temperature, salinity and pressure determine the density of seawater. Water of higher temperature and/or lower salinity is less dense than water of lower temperature and/or higher salinity so the warmest water is normally found at the surface, resulting in a stable stratification. If the water is cooled at the surface it gains density, stratification becomes unstable and the water starts sinking. If the salinity increases by evaporation or by sea ice formation water increases its density as well and sinking is enhanced. Therefore the most active sinking processes occur in high latitudes where water increases density first by cooling and then by sea ice formation. The vertical distribution of temperature and salinity in the Southern Ocean on a transect from South Africa to the Antarctic continent is displayed in Figure 5.6.

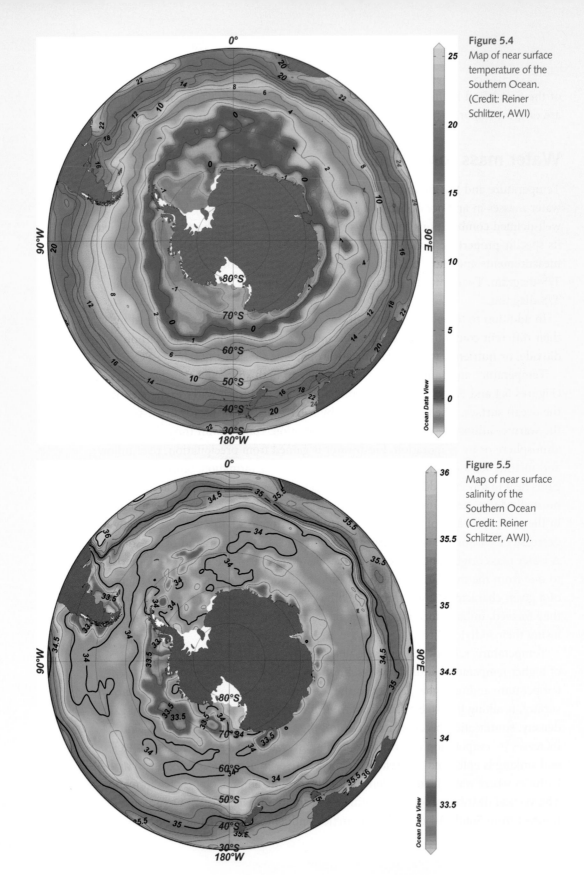

Figure 5.4
Map of near surface temperature of the Southern Ocean. (Credit: Reiner Schlitzer, AWI)

Figure 5.5
Map of near surface salinity of the Southern Ocean (Credit: Reiner Schlitzer, AWI).

Figure 5.6 Vertical transect of (a) temperature and (b) salinity on a transect from South Africa to the Antarctic continent. The major oceanographic fronts marked are the Subtropical Front (STF), Polar Front (PF), SubAntarctic Front (SAF), Southern Circumpolar Current Front (SCF), Weddell Front (WF), Continental Water Boundary (CWB) and the Antarctic Slope Front (ASF). (Credit: Eberhard Fahrbach, AWI)

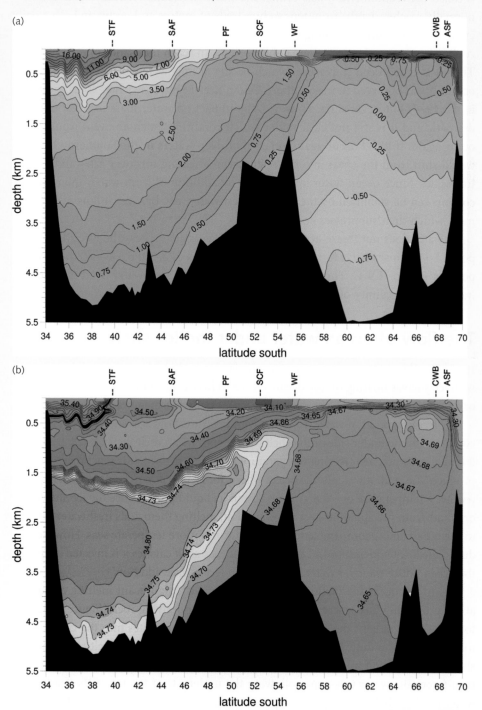

This pattern of sinking due to instability and the adjacent ascending motion is called convection, and it acts to mix the water in oceans vertically. Convection normally reaches depths from tens to hundreds of metres and occurs in both a daily cycle driven by surface cooling during the night and in an annual cycle during cooling in autumn. Under certain conditions it can reach thousands of metres depth. Then it is called deep convection which occurs at only at few places in the world ocean. It was observed in the Southern Ocean near Maud Rise where it had occurred in the 1970s but never since. However, the impact on water mass formation was dramatic and it is still not clear under which conditions such an event would occur again. If it did occur again, renewal of deep waters in the Southern Ocean would shift from the coast into the open ocean affecting turnover rates and timescales. Since ocean water loses density by addition of fresh water, the water column can be stabilised by the input of freshwater, e.g. by river inflow or ice melt, which then suppresses convection.

Seawater does not freeze at 0°C as its dissolved salts depress the freezing point. So when sea ice is formed at oceanic salinities of 3.4 to 3.5 $kg\,m^{-3}$ the freezing point is -1.8 to -1.9°C. This sea ice formation extracts freshwater from the ocean, increasing the salinity of the underlying seawater and making it denser.

The flow of ocean water under the ice shelves

In the Southern Ocean water masses can also be cooled where they are in contact with ice shelves floating on the ocean surface. This occurs either at the ice shelf front or in cavities under the ice shelf into which ocean water penetrates.

The freezing point of water decreases with pressure. So if seawater comes into contact with the under side of the shelf ice at depth it can melt the ice shelf even if it had already reached freezing temperature at the sea surface. It is this process that produces the coldest water in the ocean where, for example, at a depth of 1500 m water reaches a temperature of −2.8°C yet is still in its liquid phase. Since the ice shelves originate from precipitation over the continent, melting them adds freshwater directly to the ocean and is comparable to river inflow into more temperate seas. However, due to the low temperature and the depth of the ice shelf cavities it is injected deep into the ocean unlike the river water inflow, which is a surface addition.

Due to the ocean circulation in the ice shelf cavity, significant melting occurs at the grounding line where the shelf rests on the underlying rock (Figure 5.8). Due to a pressure gradient the relative salty ocean water penetrates into the cavity and follows the descending sea bottom to the grounding line. There, melting of the ice shelf occurs, adding freshwater to reduce the local salinity and its density, which in turn forces the water to ascend following the underside of the ice. If the fresher water reaches shallower depths the pressure decreases and the water starts

Figure 5.7 Pancake ice forms at the end of summer in March as the ocean surface begins to freeze and form winter sea ice. The exclusion of salt from the seawater during freezing generates a very strong brine which flows down to the sea floor to form Antarctic Bottom Water. (Credit: Peter Bucktrout, BAS)

freezing again. Ice platelets are formed which either ascend to the underside of the ice shelf, increasing its vertical extent, or are swept out of the cavities by ocean currents. The refrozen part of the ice shelf is called marine ice.

Icebergs are formed from glaciers reaching the coast or from the ice shelves which break off and float northwards towards the open ocean. By melting icebergs add fresh water to the ocean further offshore. With 90% of their mass underwater icebergs are at the mercy of the ocean currents. But if the sea ice cover becomes sufficiently compacted, it takes over the control of the drift, which then becomes wind driven. As long as the icebergs are in relatively cold water the melt rate is moderate but as soon as they break into smaller pieces the melt rate is significantly enhanced due to the increased surface area. Icebergs formed from marine ice attract attention after capsizing since they are of greenish colour. Icebergs can persist for several years, especially if they run aground, and during that period the ocean and the weather can produce some startling shapes and patterns.

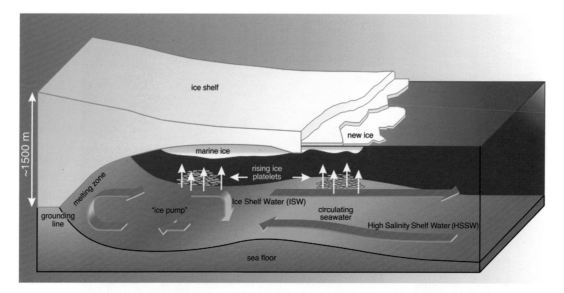

Figure 5.8 Schematic representation of the oceanic flow under the ice shelves. (Credit: Gerd Dieckmann, AWI).

The large-scale distribution of water masses in the Southern Ocean

Deep ocean currents transport the cold water away from the convective areas laterally. Since ocean currents vary in speed and direction with depth, layers are formed by water masses sandwiched between the ones above and below. This spreading of water masses can be detected on a vertical transect as a tongue of water, which can slowly mix with adjacent water masses and vary their properties.

A transect from South Africa to the Antarctic continent displays warm and saline (subtropical) water masses in the north which are limited to the south by inclined isotherms and isohalines. In mid-latitudes a tongue of fresher water, the Antarctic Intermediate Water, spreads to the north as part of a shallow circulation cell which is mainly fed by the ascending Circumpolar Deep Water. It is the densest part of the Subantarctic Mode Water that is found along the Subantarctic Front. This shallow circulation cell is of importance for the meridional heat transport and the uptake of carbon dioxide by the Southern Ocean. The Intermediate Water spreads into the subtropical and tropical oceans where it feeds the coastal and equatorial upwelling circulations. In this way the transfer from Circumpolar Deep Water to Antarctic Intermediate Water represents the upward motion of the global overturning circulation which mixes water from the top and the bottom of the water column.

A second deep circulation cell is initiated by the formation of Antarctic Bottom Water. Its origin is to a large extent in the North Atlantic. In summer the ocean around Antarctica is covered by a layer of warm, low saline surface water generated by solar radiation, precipitation and melted sea ice. In winter the water cools to freezing temperatures, sea ice forms and the salinity increases again and cold and saline Winter Water is formed. Below the Winter Water, there is a layer of warm and salty water which intrudes from the north, the Circumpolar Deep Water. Due to its high salinity this warm layer is stable below the cold, less saline Winter Water. South of and below the Circumpolar Deep Water colder deep waters are found. This bottom water originates from a flow of dense water from the shelves along the continental slope into the deep sea forming the deep circulation cell by transforming the ascending Circumpolar Deep Water in a sinking water mass which ventilates the deep ocean.

Frontal systems

The changes from one type of water mass to another do not occur steadily along a constant gradient, but in sharp steps across areas of strong gradients called fronts which are transition zones between the water masses (Figure 5.9).

Fronts were first discovered in the atmosphere as sharp separation lines between air masses of different temperatures (warm or cold fronts) and the same phenomenon was observed by oceanographers in the sea.

Fronts are of particular interest for several reasons. They are relatively simple visible indicators for the water mass distribution as the gradient is often visible at the sea surface as a change in water colour. The fronts are not only surface structures but they also separate deep water masses.

These frontal gradients in temperature and salinity produce density gradients in the ocean, which in turn induce pressure gradients that drive ocean currents. In consequence, ocean fronts are often related to enhanced ocean currents.

In addition to the flow along the front, there is motion perpendicular to it. Converging water masses produce a sinking motion, and diverging water masses produce an ascending motion, often called an upwelling. These cross-frontal motions concentrate material and plankton in the fronts and it is because of this that biological production can be so high in these areas. Deeper waters are normally higher in nutrients and frontal upwelling transports nutrients into the near surface waters. There, in the light, the higher nutrients meet concentrations of phytoplankton and biological production is enhanced. The fronts often generate stable layers, which keep the phytoplankton up near the surface and enhance primary production. This biomass feeds the zooplankton, which in turn attracts the

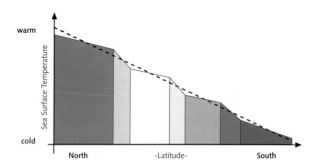

Figure 5.9 Schematic representation of water mass fronts of strong changes separating the zone of little change in water mass properties. (Credit: Eberhard Fahrbach, AWI)

whales, seals and sea birds to feed at the fronts. Such intensive life at fronts can be spotted from far away and the enhanced biological activity changes the water colour from blue to greenish.

Due to their biological impact, fronts are visible not only at the sea surface and from satellites but also in the sediments, since the increased production enhances the amount of material sinking to the ocean bottom. From this sediment record fronts can be located over geological time and since they are water mass transition zones, ocean temperature and salinity distributions can be traced back. However, since the processes are highly non-linear, the interpretation of the evidence is normally not straightforward.

The Southern Ocean off the continental slope is normally structured by five major fronts:

- the Subtropical Front at about 40°S,
- the Subpolar or Subantarctic Front at about 45°S,
- the Polar Front at about 50°S,
- the Southern Antarctic Circumpolar Current Front at about 55°S, and
- the Southern Boundary of the Antarctic Circumpolar Current, which coincides in the Atlantic sector to some extent with the Weddell Front, the northern limit of the Weddell gyre (see Figure 5.6).

The fronts are subject to intense fluctuations due to meanders and eddies. Since they are related to pressure gradients, which induce changes in sea surface height, they can be detected and monitored by satellite altimetry. Quite frequently they are also related to topographic structures on the sea floor such as ridges.

The Antarctic Circumpolar Current

The Antarctic Circumpolar Current forms a moving band around the Antarctic Continent with flow from west to east and, with a water mass transport of about 140 million m^3 s^{-1}, it is the largest current system on Earth. It is split in different

Figure 5.10 The Antarctic Circumpolar Current as derived from a numerical model (FRAM). (Credit: Southampton Oceanography Centre)

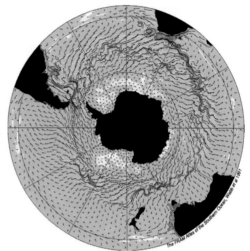

branches which are related to the oceanic fronts. The most important branches are the Subtropical, the Subantarctic and the Polar Fronts which comprise 80% of the total transport.

The Antarctic Circumpolar Current is forced along by the West Wind Drift which accelerates the surface water particles by friction. Due to the west wind increasing from north to south and decreasing then towards the Antarctic coast, near surface water transport is driven in the north by the West Wind Drift but in the south by the east wind belt. This change in transport with latitude gives rise to convergences and divergences. Convergence produces elevation of sea level, divergences a depression. Therefore in the Southern Ocean, the sea surface is inclined from north to south forming again a pressure gradient which drives, under the influence of the rotating Earth, a current to the east which spreads over the complete water column. However, ocean stratification acts through the pressure gradient related to isotherms and isohalines ascending to the south and decreases the currents towards the bottom.

The transfer of energy from the directly wind driven surface layers to the deep ocean is achieved by an intense eddy field. These eddies which are essential for the Antarctic Circumpolar Current dynamics, provide a heat transport from the north to the south across the Antarctic Circumpolar Current. This cross-frontal heat transport is an essential part of the heat budget of the Southern Ocean and plays an important role in determining the sea ice thickness.

Subpolar gyres

South of the Antarctic Circumpolar Current clockwise currents in the subpolar gyres transport Circumpolar Deep Water to the south in the vicinity of the water mass formation areas. The gyres are mainly wind-driven current systems related to the transition from the west winds in the north to the east winds in the south. The main subpolar gyres are the Weddell gyre, the Ross gyre and the Kerguelen gyre. The gyres carry source water masses to the south and newly formed water masses to the north. Their zonal structures are strongly determined by underwater topographical features. In addition to the transport of water masses, they are of importance to ecosystems, in particular, through transport of plankton and control of how long organisms spend in certain hydrographic regimes.

The role of the Southern Ocean in the global climate system

The climate system is driven by the incoming heat from the Sun and the loss of heat back to space. Net heat gain occurs in the low to mid-latitudes and net heat loss in the high latitudes. Heat has to be transported from areas of heat gain to heat loss. Poleward of 40° latitude most of the meridional heat transport occurs in the atmosphere. In the Southern Ocean significant heat transport occurs from ocean to ocean by the eastward Antarctic Circumpolar Current and by the westward Agulhas Current south of Africa. From this circumpolar heat transport a significant portion enters the South Atlantic. This surplus of heat from the other oceans results, by ocean currents, in a cross-equatorial heat transport which finally provides heat to the north Atlantic where it is transferred to the atmosphere giving rise to the mild climate of mid- and northern Europe (the Gulf Stream effect).

Water can store much more heat than air. The amount of heat contained in a 100 km high column of the atmosphere can be stored in a 3 m high water column. Because of this difference in heat capacity the ocean serves as a heat buffer for the whole global system. Up to now, 80% of the additional heat gained by the increased greenhouse effect in the atmosphere has been absorbed by the ocean. However, the largest part of the ocean cannot be heated directly due to its stable stratification. Therefore, the deeper layers of the ocean are only warmed by sinking water masses, i.e. by formation of dense water masses and convection cells. Due to the large volume of the deep ocean a temperature increase there of 0.001°C requires heat quantities which would increase the temperature of the atmosphere in the order of 1°C. Since the sinking to the deep ocean occurs in high latitudes they control the role of the ocean as a long-term heat buffer.

Finally, the heat budget of the Earth is controlled by the reflection or absorption of radiation at the surface. Bright surfaces reflect much more strongly than dark surfaces. This surface property to absorb or reflect radiation is called albedo. The albedo of ice and snow is high, whilst that for the dark ocean is low, i.e. the open ocean absorbs most of the incoming radiation and sea ice reflects most of it (Figure 5.11). If by warming, sea ice starts melting, becomes darker and finally produces open water, more heat is absorbed by the dark ocean and warming is amplified. This is a positive feedback, through which the polar amplification of change in the context of global warming can be understood. In the Southern Ocean,

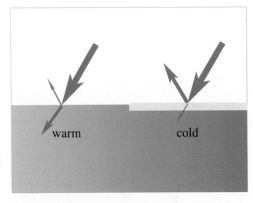

Figure 5.11 The sea ice ocean albedo effect. Bright ice reflects most of the incoming radiation, dark water absorbs most of the radiation. (Credit: Eberhard Fahrbach, AWI)

the sea ice extent is strongly influenced by heat from the ocean, since the warm Circumpolar Deep Water can reach the surface layers, if vertical mixing and convection is strong enough, whilst ocean stratification also acts on the extent, properties and duration of the sea ice through the albedo effect.

The Southern Ocean is an important area for the global carbon dioxide cycle. The formation of Antarctic Intermediate Water dissolves significant amounts of atmospheric carbon dioxide in sinking water masses. Since primary production consumes carbon dioxide and respiration generates carbon dioxide relatively old water masses like the Circumpolar Deep Water are rich in carbon dioxide. In consequence upwelling in the Antarctic Divergence areas, which is fed by Circumpolar Deep Water, leads to release of carbon dioxide to the atmosphere. However, biological activity can take up this carbon dioxide in the surface layer and extract it by sinking particles. This 'biological pump' acts to withdraw carbon dioxide from the surface layers, even if there is no sinking motion, (the 'physical pump'). The overall sum of sink and source areas in the Southern Ocean and their strength are still a matter of debate but of great significance in our efforts to predict the changes to climate over the next century.

Sea ice and polynyas

The Antarctic Ocean is covered by sea ice with a dramatic seasonal cycle. About 80% of the sea ice melts in summer. Only in very restricted areas, such as the western Weddell Sea, does multi-year sea ice occur. Sea ice is formed when ocean water freezes. First, grease ice is formed which consolidates into small floes. Under the action of the ocean waves the floes grind against each other to produce round shapes, so called pancake ice. Winds and currents, in particular tidal currents, move the floes; they merge and increase in size, and are piled up and form ridges. The dynamics of the moving ice produce a stronger growth of sea ice thickness than would be possible simply by thermodynamic growth due to heat loss alone. Thermodynamic growth is limited due to the insulating power of the sea ice. The thicker it becomes, the less heat is lost to the atmosphere, and the weaker is the growth.

Figure 5.12 In the coastal polynya off the Ekström Ice Shelf. (Credit: Frank Rödel, AWI)

Open water can form within the sea ice area allowing significant local heat loss again and a stimulation of ice growth. This diverging sea ice motion occurs in the vicinity of the continent where winds blowing offshore drive the ice away from the coast and generate open water called coastal polynyas. In these coastal polynyas open water occurs under very cold conditions, which are enhanced by the cold katabatic winds from the interior, and sea ice forms continuously here in what are considered the Antarctic 'sea ice factories'. However, the formation of leads in the sea ice due to wind and current divergences further away from the coast contribute significantly as well. A result of the intensive sea ice formation is the enhanced release of salt, which increases the salinity and density of the water in the polynyas.

Under normal conditions sea ice formation increases salinity, enhancing the water density, which in turn stimulates convection. If convection becomes intensive enough it can reach down into the levels of the Circumpolar Deep Water and initiate vertical heat transport up to the ice where it begins melting. Melting provides fresh water which in turn stabilises the water column and suppresses convection until the heat loss to the atmosphere initiates ice growth again. This freeze and melt cycle limits the thermodynamic ice growth and consequently the ice thickness of non-ridged ice to less than a metre.

Under certain conditions polynyas can be formed in the open ocean. By a combination of wind forcing in the Antarctic Divergence, vertical motion due to topographic features and vertical mixing by eddies and tides, the vertical heat transport from the Circumpolar Deep Water into the surface mixed layer can be enhanced to such an extent that the ice melts completely and open water appears away from the coast. Such conditions occurred in the Atlantic sector of the Southern Ocean at Maud Rise where an open ocean polynya was observed by satellite images around 40 years ago. The large 'Weddell Polynya' existed for several years and caused anomalous open ocean deep convection. However, since then, neither a large Weddell Polynya nor deep convection has been observed again in the vicinity of Maud Rise, but reduced sea ice conditions do occur quite often in that area.

Bottom water formation

The most important contribution to the global vertical overturning circulation from the Southern Ocean is the bottom water formation. Antarctic Bottom Water is the deepest water mass in the global ocean. It spreads up into the North Atlantic beyond 50°N.

Bottom Water formation occurs in the Weddell Sea, the Ross Sea and near Adélie Land. The Weddell Sea is the most important source since 60% of bottom water is produced there. Bottom Water is formed by dense water sinking at the continental slope into the deep sea. This dense water forms initially on the shelf where, in polynyas, huge amounts of ice form, increasing the local salinity and thus the water density. This high salinity shelf water is actually dense enough to descend the continental slope into the deep ocean, but the action of the rotating Earth tends to keep it on the shelf. Only topographic features such as canyons or eddy-related cross-frontal processes can force the dense water flow to leave the shelf and descend. Once it has reached the continental slope, the bottom water is forced to flow along the slope and sink rather slowly. In the Weddell Sea the water spilling down the slope from the southern shelf off the Filchner/Ronne Ice shelves follows the continental slope of the Antarctic Peninsula as a plume towards the northern tip, where it reaches depths below 4500 m and fills up the Weddell Basin. However, this is too deep to leave the basin since the gaps in the northern ridges are at shallower depths. Therefore, the newly formed water has either to mix with adjacent water when flowing along the slope or mix vertically in to Weddell Sea Deep Water in the basin. Weddell Sea Deep Water layers are shallow enough to leave the Weddell basin and form Antarctic Bottom Water.

Biogeochemical properties

The Southern Ocean is known to be a high nutrient–low productivity area. Normally availability of nutrients controls the productivity of an ocean area. When phytoplankton consumes nutrients and no supply occurs by upwelling of nutrient-rich deeper water, the surface layers are soon depleted of nutrients. In deeper layers there is not enough light for plants to grow. In the Southern Ocean relatively high nutrient concentrations occur in the near surface layers but still the production is low, with only some exceptions. Deep mixing, due to strong winds and weak stratification, could restrict production because it limits the time during which phytoplankton can remain in the vicinity of the surface to receive enough light for growth. More recent investigations suggest that the missing micronutrient iron limits the biological production. Iron is provided to the ocean through dust which is transported by the wind from land, e.g. from the Sahara or Patagonia into the ocean. However, in the Southern Ocean land masses which are able to provide dust are remote and, in consequence, the natural concentration of iron is very low. The hypothesis that missing iron is the origin of high nutrient–low productivity areas can be tested by controlled iron fertilisation of particular ocean areas. Recent iron fertilisation experiments support this hypothesis and suggest to some people that geo-engineering with added iron could be one way to solve the global increase in carbon dioxide.

The uptake of atmospheric carbon dioxide is an important component in the biogeochemical cycles of the Southern Ocean. Increasing carbon dioxide uptake from the atmosphere increases the acidity of the ocean waters. Increasing acidity affects biogeochemical processes. In particular, zooplankton organisms which form calcareous skeletons are affected because the solubility of calcium carbonate decreases with increasing acidity. When reaching a critical level, skeletons cannot be formed any more or will be dissolved. The potential lack of such zooplankton organisms will perturb the food web with important consequences both for the biogeochemical cycle as well as life in all the higher trophic levels. This trend seems intimately linked to global carbon dioxide levels and may now be unstoppable.

Variations and changes

The conditions in the Southern Ocean are subject to intensive fluctuations, which cover a wide range of timescales. Surface waves, small-scale turbulence and internal waves represent the high frequency part of the spectrum. Eddies and meanders with space scales from metres to hundreds of kilometres cover the timescales from days to months. Seasonal, annual, multi-annual and decadal variations form the low period part of the spectrum. Fluctuations of all timescales are superimposed to derive the ocean 'mean state' (which is only a statistical quantity and which varies with the length of the observed time series). 'Trends' are often only part of a fluctuation which is not yet evident because the duration of the measurements is too short.

The fluctuations are either induced by atmospheric or solar conditions on the oceans or result from internal dynamics such as wave formation and instabilities. The most prominent fluctuation is the annual cycle induced by changes in the solar radiation, which influences the atmospheric temperature and the winds. They induce the most dramatic change in the Southern Ocean, the build up and decay of the sea ice which extends in winter over 20 million km^2 of which 80% melts in summer. Summer warming gives rise to the relatively warm Antarctic Surface Water, which is replaced via cooling and sea ice formation by the Winter Water from which subsurface remnants endure during summer. Currents such as the Antarctic Coastal Current are subject to annual cycles due to the annual cycle of the wind and water mass formation. However, in the Southern Ocean semi-annual variations are rather widespread as well with a maximum in wind effects in spring and autumn separated by minima in summer and winter.

In the Antarctic Circumpolar Current eddies and meanders are a significant component of variability. They are mostly formed by instabilities of the currents, often in the vicinity of rough bottom topography. They play an important role in the current dynamics and in the oceanic heat transport to the south, which occurs across the Antarctic Circumpolar Current.

Multi-annual to decadal fluctuations are now interesting because time series of observations of various parameters in the ocean are getting long enough that such fluctuations are detectable. However, the short duration of oceanic observations can include the risk that interpretation as anthropogenic trends is simply part of an incomplete decadal or multi-decadal natural fluctuation. Normally atmospheric observations cover longer times than those in the oceans. They therefore allow us to detect structures such as the Antarctic Circumpolar Wave, the Southern Annular Mode (SAM) or the El Niño-Southern Oscillation (ENSO), which drive ocean change or exist as a coupled ocean–atmosphere process, which persists as changes in the ocean feedback to the atmosphere and vice versa.

The Antarctic Circumpolar Wave propagates around Antarctica inducing fluctuations in atmospheric pressure, winds, ocean temperatures and sea ice with a period of 4 years. However, it proved to be transient occurring in the mid-1980s to mid-1990s and then disappearing. The Southern Annual Mode characterises the intensity and location of the low pressure belt around Antarctica which gives rise to the West Wind Drift. Increasing SAM stands for increasing and southward shift of the West Wind Drift together with increasing temperature at the Antarctic Peninsula and decreasing ones over much of the rest of Antarctica. The SAM does not show a periodic behaviour, but seems to be random with weekly to multi-decadal timescales. There is evidence that the atmospheric circulation system over Antarctica is vertically coupled between the tropospheric circulation, i.e. the West Wind Drift, and the stratospheric polar vortex. The latter seems to be related to the stratospheric ozone concentration over Antarctica, which is strongly reduced in late winter due to anthropogenic influences, the 'ozone hole'. This suggests an anthropogenic influence on the SAM and in consequence on oceanic conditions as water mass circulation or the formation of polynyas. The extreme difference of strongly increasing air temperatures and sea ice retreat along the Antarctic Peninsula and the cooling and sea increase in the East Antarctic can be explained by the change of the SAM to a more positive state.

The third mode of variability is related to ENSO via the Pacific American mode (PSA) which seems to reach from the Pacific into the Southern Ocean influencing wind, sea ice and water mass properties, to an extent that ENSO signals can be detected in animal growth in the Weddell Sea. The Pacific and the Atlantic Sectors are often in an opposite phase, i.e. shrinking sea ice in the Bellingshausen/ Amundsen Seas correlates with increasing sea ice cover in the Weddell Sea. This feature is often referred to as the Antarctic Dipole.

The superposition of these different modes of variability does not yet allow a clear understanding of the effects of climate change on the oceanic conditions in the Southern Ocean but models suggest a relationship to the trends in SAM and increasing greenhouse gases as well as depleting stratospheric ozone.

Observation methods

Observations in polar oceans present particular challenges. On the one hand, the harsh conditions make procedures and methods developed for mid-latitude oceans very difficult, whilst on the other hand, as temperature varies by an order of magnitude less in polar oceans than in lower latitudes, at least the same accuracy of measurements is required to detect variations.

Typically ice strengthened research vessels or even icebreakers are used to deploy a range of instruments. One of the most common instruments is the CTD probe measuring conductivity, temperature and depth combined with a system of water samplers. The depth of the measurements is obtained from a pressure sensor whilst the salinity is derived from electrical conductivity, temperature and pressure. All parameters are transformed into electrical signals transmitted back to the ship through a conductor cable, which also supplies power and carries water bottles to collect samples for further analysis of concentrations of nutrients and trace substances at specific depths. Temperature measurements require an accuracy of $0.001°C$ and salinity of $0.001\,\mathrm{g\,kg^{-1}}$ to detect variability in the interior of the

Figure 5.13 German research vessel *Polarstern* in the ice of the Weddell Sea. Built with ice-breaking capability, this is one of the few research vessels that can operate in winter ice. (Credit: Tony North, BAS)

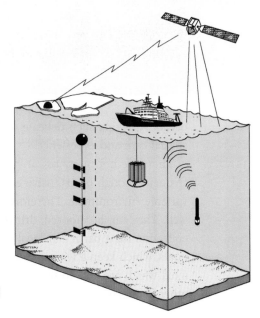

Figure 5.14 Oceanographic observation methods include satellites, sensors lowered from a ship, buoys and sensor arrays fixed to the seabed. (Credit: Eberhard Fahrbach, AWI)

Figure 5.15 A temperature/conductivity/ depth probe (CTD) with a series of water samplers is deployed to provide vertical profiles for physical and chemical parameters. (Credit: Peter Bucktrout, BAS)

ocean. Applying such delicate sensors under freezing conditions requires special care and experience.

There are not many research vessels, so ship-borne measurements are not frequent enough or extensive enough to characterise comprehensively the highly variable environment. Therefore autonomous observation platforms are used as well. They can be moored, drifting or self-propelled. Moorings provide data collection for a particular locality and consist of a ground weight and buoyancy floats, which keep a wire upright in the water column. On the wire, instruments are fixed which are able to measure time series of temperature, salinity, currents and other variables. The data are stored in a memory and can be retrieved after the mooring is recovered. Special sensors, upward-looking sonars, measure the sea ice thickness by an echosounder, which determines the distance from the instrument to the ice underside. Sinking particles can be collected with sediment traps and the vocalisations of marine mammals can also be recorded. Bottom pressure recorders indicate variations of the sea surface height and in combination with inverted echosounders the change of water mass properties.

Drifting instruments, which transmit their data and positions, can be applied at the sea surface or deployed on ice floes and move with the water masses. Vertically profiling floats measure profiles of temperature and salinity every couple of days, when they leave their parking depth at about 1000 m, and descend to 2000 m depth. Normally every 10 days they come to the surface where they use satellite communication to determine their position and to transmit the recorded data. If the ocean is covered by sea ice, they cannot reach the surface and they avoid hitting the ice by detecting when the temperature is close to freezing and terminating the ascent. Data are stored until the float reaches the surface the next time.

Recently marine mammals, e.g. elephant seals, have been used to carry sensors, which transmit their data via satellites. The animals migrate over long distances and reach areas that are difficult to reach by ship or buoy. Since they need air to breathe they search actively for open water and in consequence are able to transmit data rather frequently. The instruments provide not only the physical environmental data, but also tell us about the foraging behaviour of the animals and their physiological performance.

Whereas *in-situ* instruments can only cover a very limited area and time period, satellites have the potential to obtain regional data rather rapidly. Data now include sea surface temperature, salinity and colour, roughness (from which wind and waves can be derived) and sea surface height (which also allows ocean currents to be determined). Satellites can also measure sea ice concentration and extent, sea ice drift and to some extent sea ice thickness.

The combination of satellites and *in-situ* data allows an increasingly detailed description of the present state of the Southern Ocean and the detection of changes, but compared with other oceans the data from the Southern Ocean are still very limited. The new international initiative Southern Ocean Observing System is intended to collect many types of data from all over the Southern Ocean throughout the year. Significant efforts are needed to obtain data sets that are sufficiently dense in space and time to support or reject hypotheses on changes. Here, numerical models can help to fill the gaps by covering the Southern Ocean and its variability over longer times and on a complete grid. Since they are based on physical laws, they can give answers to the question of the origin of the observed fluctuations and help to predict future development.

Conclusions

The Southern Ocean is a major part of the global climate system. Due to the harsh natural conditions it is, in spite of a century of exploration, still poorly known.

It links the three major ocean basins together through the Antarctic Circumpolar Current, the largest ocean current on Earth, forming a global oceanic system. It plays a dominant role in the global ocean overturning circulation by upwelling water masses from the deep and afterwards sinking back into intermediate depths and to the bottom, filling the global oceans' bottom layers. The overturning circulation affects storage of heat and gases, with oxygen and carbon dioxide solubility acting as a buffer for atmospheric changes whilst the biogeochemical conditions such as oxygen content and acidity directly influence the living conditions for many organisms, in particular in the deep ocean. Interrelation between physical and biochemical processes, as well as potential feedback loops such as the dependence of carbon dioxide uptake on storminess, present particular scientific challenges. The wide range of variability for many measurements in both space and time makes it difficult to detect anthropogenic change clearly and to distinguish it from natural variability. Therefore the greatest care is needed in interpreting model predictions for the future, especially for warming and acidification scenarios.

6 | Life in a cold environment

PETER CONVEY, ANGELIKA BRANDT AND STEVE NICOL

The ship was stopped to take soundings and to dredge, and a good haul
was brought up from the bottom of the most beautiful things imaginable.
The bottom of this sea must be covered with the most wonderful collection
of starfish, sea urchins, Polyzoa, crustaceans, etc. All fairy-like delicate
things that reminded one much more of the Tropics but all living at a
temperature of 29°F. **Edward Wilson, *Diary of the* Discovery *Expedition***

Antarctica is a good place to study both evolution and adaptation. Despite the low
temperatures and lack of liquid water on land, and the effects of sea ice and
temperatures just above freezing in the sea, Antarctic organisms have found ways
not only to survive but to thrive in what to us seem very difficult environmental
conditions. Its unusual evolutionary history, from Gondwana supercontinent to the
present day, together with a climate that has now been cold for at least 25 million
years, gives Antarctic biodiversity some unique features, whilst the survival
techniques that its plants, animals and microbes have developed can provide
important insights into the physiological and biochemical ways in which organisms
function. It is on land that physical environmental extremes are perhaps most
obvious, and the continent's terrestrial habitats include some of the coldest, driest,
windiest and most abrasive to be found anywhere on Earth. Less than 0.4% of the
Antarctic is ever free of ice or snow cover, and even then this may be for as little as a
few days or weeks during the height of the summer. In this respect Antarctica is
strikingly different to the Arctic. The latter experiences a similar radiation
(daylength) regime, determined by latitude, which means that both polar regions
face similar thermal stresses during their winter months when the input of solar
energy is absent. In summer, the almost complete ice cover of Antarctica gives it a
very high albedo, reflecting virtually all incoming solar radiation back up through

Antarctica: Global Science from a Frozen Continent, ed. David W. H. Walton. Published by Cambridge University
Press. © Cambridge University Press 2013.

the atmosphere. In contrast, large areas of the northern continents and archipelagos, and, increasingly, the surface of the Arctic Ocean, become ice free for several months during summer, enabling the absorption of solar radiation and leading to temperatures several degrees higher on average than at comparable Antarctic latitudes, sufficient to support considerable biological productivity.

The Antarctic marine environment also faces environmental extremes. As well as being one of the coldest, it is also one of the most thermally stable marine habitats on the planet, and faces considerable disturbance from various forms of ice. As the surface layer of two-thirds of the Southern Ocean freezes during the winter, the 'area' of the southern hemisphere covered by ice can be doubled. The extreme seasonality of the solar regime that underlies this also has the biological consequence of generating short pulses of primary productivity during the summer months, and therefore intense seasonality of food supply is an important challenge faced by consumers. Sea ice and anchor ice formation in shallow water directly impact benthic and littoral/intertidal habitats, though this typically only influences the first few metres of depth. However, glaciers and ice shelves calving directly into the sea generate icebergs of widely varying size and depth, whose movement can cause considerable damage and destruction to benthic habitats, frequently to depths of tens or hundreds of metres, and occasionally to more than 1000 m.

Even without taking the extent of winter sea ice into account, the continent of Antarctica is larger in area than the United States, Australia or Western Europe. It would be extremely simplistic to think that it can be treated as a single entity in biological terms. In describing how organisms have managed to make a success of life in this distant and apparently hostile region, we have started with an initial division into three distinct environments – the marine pelagic and benthic ecosystems and the terrestrial ecosystem – recognising, however, that there are many cross-cutting themes and linkages that provide some of the most exciting of contemporary research fields.

Life in the pelagic zone

The pelagic zone of all oceans is, in terms of volume, the largest ecological zone (biome) on the planet, reaching from the water's surface to the ocean depths and extending from the coastal zones of all the continents. It is in the upper reaches of this zone where sunlight stimulates growth in the unicellular plants, the phytoplankton, and this starts the food chains that eventually lead to the seabirds, whales, fish and to humans. Sunlight only reaches into the top 200 m or so of the ocean and, below this productive layer can lie several kilometres of ocean that, other than some exceptional communities based on chemosynthetic processes, is dependent on what sinks or swims out of the surface layer to support the food webs living in the permanent darkness of the abyss. In the Southern Ocean, little is known about the animals that swim in the deep ocean but there has been a surprising amount of study of the plants and animals that inhabit the surface layers and we will concentrate here on what is known about the ecosystems that occupy the sunlit layer of the waters that surround Antarctica.

Life in the ocean is dependent on light from above. Sunlight gets absorbed very rapidly by water, and by the particulate and dissolved material, so there is a gradient of light from the surface into the deep. In the Southern Ocean, there is another factor that comes into play – sea ice. During autumn, the surface of the ocean freezes and a 1–3 m thick layer of sea ice extends over the most productive parts covering an area of up to 22 million km^2 – roughly the same size as the Antarctic continent itself. Sea ice has a number of effects on the Southern Ocean. As seawater freezes as a result of being in contact with the very cold air, the ice formed is fresher than the ocean and the salt that is extruded in the freezing process makes the underlying water very cold, saline and, hence, very dense. This water sinks down to the ocean floor and seeps through the ocean basins of the world, coming to the surface in the northern hemisphere many centuries later. The formation of this 'Antarctic Bottom Water' is an important driving element of what has become known as the 'global conveyor belt' of globally interlinked ocean current systems.

Because of the sinking of this dense water there has to be a complementary rising of water to compensate, and this water which upwells around the Antarctic continent is rich in nutrients and is thought to be one of the reasons why some areas of the Southern Ocean are so highly productive. Deeper waters are richer in nutrients because there is a tendency for all the plant and animal production in the surface layer to sink and the nutrients that are incorporated into biological particles – cells, waste material, dead bodies – to fall out of the sunlit zone and into the abyss. This strips the nutrients out of the surface layer and they have to be replenished where upwelling occurs. Upwelling is vital because in summer the surface of the ocean warms in the sunlight and can become thermally isolated from the deeper layer, thus preventing vertical movement of water and nutrients. Although ice has a positive effect on plant growth by initiating the upwelling circulation, it also has a negative effect by cutting off light from the surface layer as it grows in autumn and winter. The amount of light that is absorbed by the ice depends on its thickness and the amount of snow that falls on top of it, but ice also forms a substrate on which algae can grow. The algal communities that develop over winter can colour the ice brown and can absorb significant amounts of light, as well as supporting a considerable faunal community including, for instance, being one of the main nursery grounds for krill, one of the 'keystone' species of open water pelagic ecosystems in the Southern Ocean. This sea ice ecosystem is one of the features that makes the Southern Ocean unique.

Box 6.1 **Sea ice: an unexpectedly productive habitat**

Research in recent years has indicated that sea ice plays a crucial role in the seasonal cycle of production of the Antarctic marine ecosystem. As ice forms in autumn, some of the algae and other microbes in the water become incorporated in the crystallising ice and thrive in the complex habitat of the brine channels which permeate it. The ice floating on top of the ocean keeps these plant communities as close to the sunlight as possible and during the dark months this is the best place to be to encourage growth. In spring with longer hours of sunlight, the sea ice communities begin to develop and they can then become the gardens on which the ocean's herbivores can graze. During winter there is little food available in the water column for the herbivores to feed upon so they employ many strategies to cope with this long, dark period of food shortage. Some species hibernate, some can retreat to the ocean floor where the remains of the summer production may lie, some starve and some rely on supplies laid down over summer. Adult krill (Figure 6.1c) can probably adopt a range of these strategies but larval krill, which are hatched in late summer, need food to allow them to grow and develop over winter. For them, the sea ice ecosystem is probably the critical food source and it has been suggested that there is a strong relationship between the amount of sea ice in winter and the number of larvae that survive into spring.

When the sea ice melts in spring it releases into the water all the organisms that have been growing over the winter, and these are thought to be the seeds that initiate the bloom of primary productivity. The melting sea ice also affects the physical structure of the ocean because the melt water is much fresher than seawater and it tends to form a less dense layer, separated from and floating on the underlying ocean. The algae thus tend to be circulated within this shallow surface layer and are kept in the sunlit zone, which further enhances their growth. As the sea ice retreats from north to south it leaves in its wake a bloom of phytoplankton and it is this intense burst of productivity that provides the base of the food chain that eventually leads to the great whales, and even to humans.

(a)

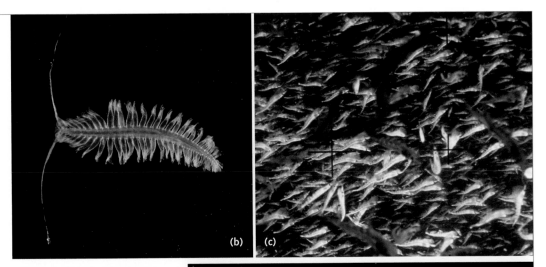

Figure 6.1. Pelagic animals in the Southern Ocean water column. (a) Jellyfish under sea ice. (Credit: Simon Brockington, BAS); (b) A polychaete (*Tomopteris* sp.) from the water column above the Bellingshausen Sea continental shelf. These worms swim with their wide fleshy parapodia (like paddles) and are widely found around Antarctica. (Credit: Peter Bucktrout, BAS); (c) Krill (*Euphausia superba*) are small, shrimp-like animals that grow up to about 6 cm in length and live for up to 5 years. Krill usually live in dense swarms, which may have more than 10 000 krill in each cubic metre of water. (Credit: Jon Watkins, BAS); (d) A swimming sea cucumber from 500 m depth on the Bellingshausen Sea continental shelf. Remarkably some individuals of this species were found within 100 m of the surface around 400 m above the seabed. This species is quite transparent and the details of its digestive system can be seen. (Credit: Peter Bucktrout, BAS); (e) *Notothenia* amongst Bryozoan covered rocks 39 m underwater. (Credit: David Barnes, BAS)

Figure 6.2 The blood of icefish (Channichthyidae) is transparent as it lacks haemoglobin and so the fish must absorb oxygen directly through their scale-less skin. (Credit: Peter Bucktrout, BAS)

The Southern Ocean hosts several pelagic ecosystems whose borders are defined by a series of physical features. Close to the coast is a community that is most affected by ice and which has the greatest connection between the surface waters and the seafloor. The Antarctic continental shelf is unusually deep (typically ~500+ m) because of the weight of the ice sheet pressing down on the continent itself. None the less, there are a number of species, such as Antarctic silverfish and the coastal ice krill *Euphausia crystallorophias*, which can be found throughout the water column, in association with the benthos and even deep beneath the great ice shelves. There is a fairly diverse fish fauna on the shelves and on the continental slope, which includes numerous members of the notothenoids, an endemic order of Antarctic fish which are characterised by their lack of swim bladder. They include, in the icefish, the only vertebrates that completely lack haemoglobin in their circulatory system and myoglobin in their muscle tissue. The Antarctic silverfish (*Pleuragramma antarcticum*) is a schooling species that resembles the more northern herrings or anchovies and which can, at times, form an important element in the diet of emperor or Adélie penguins. The ice krill is another species that can be important food for vertebrate predators including minke whales, seals and penguins. The distribution of both the silverfish and the ice krill cuts out sharply at the edge of the shelf break where a strong westward flowing current jet forms a demarcation between the coastal zone and the more offshore communities. The phytoplankton communities of the coastal zone are less distinct than their animal counterparts but the productivity of this zone is the highest of the region, in the short period when the continental shelf is ice free. Vast blooms of algae such as *Phaeocystis* can turn the waters intensely green and the detritus from these blooms sinks to the seafloor feeding the benthic communities there.

Immediately offshore of the shelf break lie the highest concentrations of krill which, not coincidentally, are still within the foraging range of the seals and penguins that breed on the continent (or on Antarctic or sub-Antarctic islands). The zone dominated by krill may extend for several hundred kilometres, as in the southeast Atlantic, or may be closely restricted to the coast, as it is in the southeast Indian Ocean. There is an intimate relationship between the distribution of krill and the physical environment – particularly the currents and the distribution of sea ice, which is itself affected by the currents. The two current systems of the Southern Ocean – the broad Antarctic Circumpolar current (ACC) and the narrower, westwards flowing, coastal current interact in a series of gyres, whose location is dictated by the submarine topography and by the shape of the coastline. The frontal systems that delineate these currents can also form the borders to oceanic communities. Between the two currents lie the Southern Boundary of the ACC and the Southern ACC Front, and these features probably mark the northernmost extent of the krill-based ecosystem in most regions of Antarctica. They often coincide with the maximum extent of sea ice in winter. Not surprisingly, the majority of krill-feeding vertebrates are found south of these fronts.

Krill are not the only oceanic herbivore, and representatives of most of the common worldwide groups of pelagic organisms are to be found in the waters that surround Antarctica. Krill compete with – and probably consume – copepods, gelatinous zooplankton, worms, molluscs and microscopic consumers. The diversity of organisms in the pelagic zone is actually lower than is found in other productive upwelling zones of the ocean and many species are slow growing and longer lived.

Birds

The most studied species in the Antarctic are undoubtedly the penguins, of which four species (emperor, Adélie, chinstrap and gentoo) breed on and around the continent and its nearby islands, with other species such as the king, macaroni and royal penguin on the sub-Antarctic islands. The bird populations in Antarctica number many millions with many of them dependent on krill and other oceanic prey for their principal food. This makes seabirds a good indicator of changes in the marine ecosystem, and research programmes exist which monitor variability and change in seabird populations that can be associated with, for example, fisheries activities and environmental change. There is evidence that more temperate species, such as Gentoo penguins, are breeding further south along the Antarctic Peninsula than previously, and numbers of the truly cold-adapted penguins – Adélie and emperor – may well be declining in the northern parts of their range. Of the eight petrel species found, four – Antarctic petrel, snow petrel, southern giant petrel and Wilson's storm-petrel – breed on the continent whilst the remainder breed in

Box 6.2 **The essential link: krill**

Within the pack ice zone one animal is dominant, though it is rarely as visible as the animals that depend on it. This animal is krill, a pelagic shrimp-like crustacean that occurs in abundance in south polar waters. Krill are relatively large (up to 6 cm long) for pelagic herbivores and are highly nutrient rich, and these two features, together with their aggregating habit and their huge abundance in some localities, make them the preferred food for most of the vertebrate species of the region – the baleen whales, seabirds, seals and fish. Once thought to be short-lived animals, they are now known to be able to live to 11 years old, which makes them considerably older than many of the animals that prey upon them. Krill live in schools or swarms that can extend for many kilometres and contain millions of tonnes of crustacean biomass. Not surprisingly, they have been commercially fished for the last 40 years, with a current annual catch of 210 000 tonnes being taken from the South Atlantic. The fishery appears to be in an expansion phase following a period of stable catches. The catch is used as aquaculture feed, and for a variety of medical and pharmaceutical uses, including the production of omega 3 oils for human consumption. The global demand for farmed fish may drive the krill fishery into more rapid expansion and this will require careful management to ensure the needs of other elements of the ecosystem (see Chapter 10). Not only are krill vitally important for the wellbeing of the vertebrates of the Antarctic region, they also exert a major influence on the microscopic organisms in the water as the huge schools move around and strain out most of the particulate organic matter. It has also recently been discovered that krill may regularly migrate and presumably forage across great depth ranges of several thousand metres. There are suggestions that the waste produced by krill schools fertilises the ocean and encourages phytoplankton growth once they have moved on.

the sub-Antarctic. In addition there are cormorants, skuas, gulls and terns, as well as sheathbills, the latter the only Antarctic birds without webbed feet. Seabirds forage in the surface of the water with the smaller species, such as the snow petrel, dipping their prey from the surface film, larger flying birds plunging into the top 10 m or so and the penguins diving to great depths (up to 50 m) in search of their prey, which consists of krill, squid and fish. During their breeding seasons, birds are tied to the colonies to which they have to return regularly with food for their offspring, whereas in winter they can disperse more widely. Modern studies using satellite trackers have demonstrated that birds such as Adélie penguins can cover thousands of kilometres on their winter feeding trips, but return faithfully to their natal colonies in spring in time to breed.

Box 6.3 **Emperor penguin**

Figure 6.3 Emperor penguin, the largest penguin and the only one to incubate its eggs through the winter. (Credit: Angelika Brandt)

The largest of the living penguins (fossils of a species almost twice as tall have been found), emperor adults are over 1 m tall and weigh 20–40 kg. They breed during the winter on the sea ice in what must be one of the most inhospitable habitats on Earth. The size of the total population is not known with any accuracy as many of the colonies have rarely been visited, but it is estimated that there may be 400 000 birds and, recently, remote-sensing techniques have been applied to provide a more accurate overall population estimate. After laying a single egg in May, which the male then balances on his feet and keeps covered with a flap of skin, the female leaves for the long walk to the sea across the ice and 2 months' feeding. Meanwhile the male fasts until the egg hatches around 8 weeks later and the female returns to feed the chick and take over parental duties. Then, at last, the male can walk up to 100 km to the sea to feed. Both during the incubation and the young chick stage all the adults form a dynamic huddle through which every bird slowly walks, gaining an equal share of protection from the low winter temperatures and high winds. Because most emperor colonies are found on fast ice they are likely to be a species that is directly affected by changes in the sea ice environment that may occur as a result of global warming.

Box 6.4 **Albatrosses**

Figure 6.4 A wandering albatross, the largest sea bird in the world. (Credit: NOAA)

The wandering albatross is the largest seabird in the world, with a wing span up to 3.35 m, and it holds a strong place in our cultural history (Figure 6.4). Gliding on the endless wind they can cover up to 800 km in a day. On the sub-Antarctic islands there are also grey-headed, black-browed, light-mantled sooty and royal albatrosses. Killing these birds is reputed to bring bad luck, yet hundreds of thousands of albatrosses have been killed in the Southern Ocean fisheries over the last few decades, caught on long-line hooks or drowned in gill nets (Figure 6.5). As they spend most of their lives at sea, it was not until the development of microelectronic logging systems that scientists were able to track where they went when they left the nest. The tracks show how the younger ones circle the continent whilst those older ones, those with chicks, may travel down to the Weddell Sea or up to the coast of Brazil or Africa to find food (Figure 6.6).

Figure 6.5 A regurgitated long-line fisheries hook and plastic bags alongside a wandering albatross nest site at Bottom Meadows, Bird Island, South Georgia. (Credit: Peter Bucktrout, BAS)

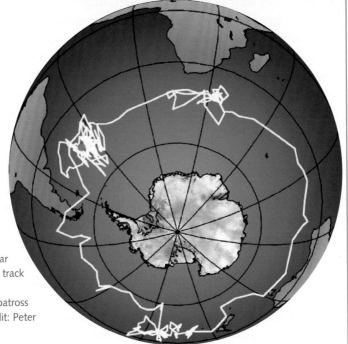

Figure 6.6 A geolocator is a miniature device attached to birds which signals its position at regular intervals to a satellite. This track shows the long-distance circumpolar flight of an albatross from South Georgia. (Credit: Peter Bucktrout, BAS)

All albatross species are long-lived (possibly over 50 years), delay their breeding until they are 7–10 years old and rear only a single chick, making their populations very vulnerable to mortality caused by fisheries. Since all populations are declining, some very rapidly, international activities have focussed on mitigating the deaths by promoting changes in fishing practices and, although this has had a positive effect for licensed trawlers, unlicensed pirate boats continue to kill many birds.

Figure 6.7 Penguin guano, the distinctive colour is due to the high krill content. (Credit: Steve Nicol)

Flying seabirds can range over great distances in the Southern Ocean but still have to return to colonies on the continent or on islands to breed. Unlike the seals and penguins, flying seabirds are restricted to feeding in the very surface layer of the ocean and some species, such as snow petrels, are adapted to feeding in the leads between the floes in the disintegrating pack ice. Some, such as sooty shearwaters, migrate out of the region to breed, using the rich waters of the sea ice zone in summer merely as a fattening zone.

Life on the Antarctic continent occurs largely around the coastline and this is because the nutrients that support most forms of Antarctic life arise from the surrounding ocean. The most highly visible signs of life on land, the penguin and seal colonies, are entirely dependent on the pelagic food web for their diet, and their guano provides the major nutrient source for many terrestrial food chains.

Seals

Colonies of fur seals and Weddell seals face the same constraints as the land-breeding penguins, although fur seals are generally found breeding in the north of the region where persistent ice is less of a problem, while Weddell seals are able to maintain holes in the ice so that they can get to their food (mainly fish and other benthic organisms) despite the presence of unbroken fast ice. The seals of the pack ice – the Ross and the crabeater – are thought to spend their entire lives in the sea ice zone using the ice as a platform in winter from which they can access the concentrations of over-wintering krill, fish and squid that persist over the continental shelf and slope. Fur, Weddell and elephant seals, being larger animals, which return to colonies to breed, present opportunities to study the feeding behaviour of pelagic predators. Using satellite trackers and a variety of advanced electronics, it has been possible to determine the foraging behaviour of these wide-ranging animals. Elephant seals have been shown to dive to great depths (at least 1.5 km) in search of fish and squid, ranging in latitude from temperate waters to the frigid sea ice zone. Fur seals regularly travel from the sub-Antarctic where they breed to the ice edge where they feed. Crabeaters are thought to be the most numerous seal on the planet because of their extensive habitat amongst the circumpolar ring of sea ice. Recent results from a global census of pack ice seals

Figure 6.8 Male Antarctic fur seal on the coast of South Georgia. (Credit: Peter Convey, BAS)

Figure 6.9 A crabeater seal on an ice floe near the Antarctic Peninsula. Crabeaters are the most numerous seals in the world, with a population over 10 million. While they do not eat crabs, they do eat krill and other crustaceans. (Credit: Jeffrey Keitzmann, NSF)

Figure 6.10 Humpback whale. Once heavily fished, populations in the Southern Ocean are slowly recovering. (Credit: Angelika Brandt)

suggest that their numbers are relatively stable and their abundance is probably in the order of 10 million animals, somewhat lower than earlier estimates. Thus it seems unlikely that crabeater seals have increased greatly in numbers in response to an increase in food supply caused by the demise of the great whales.

Whales

Historically the most important higher order predators in the Southern Ocean have been the great whales. Whilst in the mind of the public the Antarctic is generally associated with whales, most species are only there in the summer, migrating south in October/November to graze on krill and phytoplankton as the sea ice melts. Southern Ocean whales can be divided into those that feed predominantly on krill (blue, fin, humpback and minke), those that feed on other small pelagic species such as copepods (right whale), and the toothed sperm whales that are squid and fish eaters. Killer whales are also important predators of other whales, seals and penguins. Most of the krill-feeding whales are thought to migrate north in autumn to breed in subtropical or tropical waters. There is, however, some evidence that minke whales may remain in the pack ice zone throughout the year, and recent research, using acoustic recording devices which locate whales by their songs, has revealed that a proportion of the populations of other great whales may also remain longer in Antarctic waters than was first thought.

There appears to be some segregation of the oceanic habitat by the baleen whales, with the minkes being most associated with the sea ice, while blue and humpback whales are mostly found offshore of the ice edge, and fin whales farther offshore still. This separation means that these four species are thought to consume different size classes or species of krill, which also exhibit a degree of spatial segregation with ice krill being found inshore or in the deep bays, juvenile

Antarctic krill being closer inshore and adults occurring over deep water. The amount of krill that must have been consumed by baleen whales in the years before commercial whaling took its toll was prodigious – some estimates put it as high as 200 million tonnes a year – and this still left enough food available for large populations of seabirds, seals and fish.

Whilst all populations were reduced by whaling in the twentieth century, the suspension of commercial whaling in the Southern Ocean since 1984 has given some species an opportunity to recover. Some populations of humpbacks are certainly recovering as apparently are southern right whales, but the information for fin whales is presently too poor to say with any certainty what their population trend is. Minke whales have remained present in considerable numbers and are now the main target for Japanese scientific whaling, with a catch of a few

Box 6.5 The astounding productivity of blue whales

The blue whale is the largest animal that has ever lived and can grow to a weight of 140 tonnes and a length of 35 m. Blue whales reach maturity at 5 years, they have a 10–12 month gestation period and produce a 2.5 tonne calf every 2–3 years. All of this is supported on a diet of pure krill, and during the Antarctic summer an individual blue whale may eat 40 million krill a day (around 3.6 tonnes). This makes them especially vulnerable to any direct effects of climate change on krill availability. In comparison, the largest land animal, the African elephant, which can weigh 12 tonnes, reaches maturity at 9–12 years, has a 22 month gestation period and produces a 120 kg calf every 5 years. The prodigious productivity of blue whales is even more amazing because they are generally thought to confine their feeding to the 6–8 months of the year when they are in the Southern Ocean. Obviously, the pre-human-era food web of the Antarctic region was sufficient to support not only the huge populations of blue whales that existed before exploitation but also large populations of other krill-feeding whales such as fin, humpback and minke, and enormous populations of krill-dependent seals, seabirds and fish. About 1.3 million whales were killed during the years of commercial whaling in the Antarctic, including 350 000 blues; the blue whale population remains below 15% of its pre-exploitation level, despite 40 years of protection, and their numbers are so low that it is difficult to estimate a population size. Scientists are still debating the effect on the ecosystem of the removal of such a high proportion of the baleen whales and whether they will ever be able to recover. Some even speculate that the ecosystem may have changed forever and there will be no return to the days when 'Whales' backs and beaks were seen at close intervals quite near the ship, and from horizon to horizon ... The sea was swarming with *Euphausia*' (Bruce, 1915).

Figure 6.11 Leopard seals are important predators, eating penguins, fish and krill, and have been known to kill other seals, especially young crabeaters. (Credit: Peter Rejcek, NSF)

hundred each year. Because minke whales remain within the pack ice for much of the year they are difficult to count from ships and there is therefore considerable debate over how large their population is. Current non-lethal research on Southern Ocean whales is concentrating on their role in the ecosystem, with some studies tracking whale migrations using electronic tags, others using molecular techniques to identify food from faeces samples and to examine population structure, and others looking at populations by identifying their underwater 'songs'.

The true top predators of the Antarctic region, apart from humans of course, are the orcas, the leopard seals and seabirds such as giant petrels and skuas. These vertebrates prey on other large animals of the Southern Ocean, orcas taking a range of prey items from fish to the young of great whales, and leopard seals concentrating on smaller food items such as penguins, krill and fish. The predatory birds haunt penguin rookeries and other bird colonies in search of eggs and chicks.

The open ocean

Once out of the reach of winter pack ice, the marine ecosystem begins to resemble the oceanic ecosystems of other parts of the globe. The vast open ocean is stormy and relatively unproductive, despite there being significant quantities of nutrients. The food chains here support populations of microscopic animals and plants as well as smaller invertebrates, but these are scant pickings for the larger species such as whales and seabirds. Dotted around the Southern Ocean are islands or island groups, and, where the circumpolar current flows over the shelves surrounding these islands, it brings up deep nutrients that stimulate localised blooms that stream out around the islands, especially in the lee, making them biological hot spots. The islands also provide rare and globally important breeding places for seals and seabirds which can often be crowded into vast colonies, fed by the productive waters surrounding them. A further measure of the exceptional

Figure 6.12 Satellite image of phytoplankton bloom off South Georgia. (Credit: NASA)

local productivity associated with these islands is provided by the early history of the onshore whaling industry of South Georgia in the South Atlantic – here, it took several years after the establishment of the onshore processing facilities, before whale catching boats even had to venture out of the sheltered bays of the

Figure 6.13 Rookery of gentoo penguins. (Credit: Steve Nicol)

island's north-east coast in order not to be limited by the numbers of whales available to them.

The pelagic ecosystem of the Southern Ocean has experienced vast changes over the last 200 years – much of it induced by humans. First the massive sub-Antarctic island populations of fur seals were systematically hunted almost to the point of extinction. Elephant seals and the great whales were next and some of their populations are still recovering, whilst others remain precarious. Fish populations were the next target, and the large populations of the few commercially attractive species were rapidly reduced to low levels in the 1970s, and they remain there, leaving only two fish species (toothfish and icefish) that are currently harvested commercially at relatively low tonnages. Finally, krill has been harvested for the last 40 years, although the current level of catch is low (~210 000 tonnes per year) relative to the estimated biomass, and management measures are now in place to attempt to avoid the mistakes of the past (see Chapter 10). The harvesting of the animals of the region has also had unintended consequences. For example, many tens of thousands of seabirds have been accidentally killed during long-line fishing operations in the Southern Ocean and this 'incidental mortality' has reduced several species of albatrosses and petrels to extremely low levels, including the very real threat of extinction. Contrast this effect with the anecdotal accounts of the vast numbers of seabirds that must have been supported by the offal discards from the industrial whaling operations of the past and it becomes evident that human interference in the ecosystems of the Antarctic has a long history and that the effects of this interference are not always predictable.

Marine benthic ecosystems

Estimations of the biodiversity of the Antarctic continental shelf are still inadequate as many geographic areas are still to be sampled representatively (e.g. those of Wilkes Land and the Bellingshausen and Amundsen Seas), although the outcomes of the recent Census of Antarctic Marine Life are leading to considerable increases in coverage. The benthic groups with the most species are the poriferans, polychaetes, gastropods, bryozoans, amphipods, isopods and ascidians and, in some specific locations, also bivalves and gastropods. Sessile taxa are favoured on the continental shelf in the Southern Ocean, possibly because of the preponderance of substrata provided by glacial-marine sediments and dropstones. Epifaunal taxa appear to have adapted well to the coarse-grained glacial substrates, with sessile filter and particle feeders being especially prominent. Dense communities of sponges, ascidians, anemones, hydroids, gorgonians, corals, bryozoans, and crinoids are characteristic of the modern Antarctic shelf fauna at depths below the zone of the influence of ice scour and anchor ice formation. This fauna often forms stratified three-dimensional communities (analogous with tropical rain forests). Associated with this sessile biota are mobile and wandering or even swimming taxa such as echinoderms (ophiuroids, asteroids, echinoids, holothurians), pycnogonids, isopods, amphipods, nemerteans and gastropods. There are two major types of shelf communities. One is dominated by sessile

Figure 6.14 Sponge community on the Southern Ocean shelf. (Credit: Julian Gutt, AWI)

Figure 6.15 Benthic organisms on the seabed of the Southern Ocean. **(a)** A sea anemone using a sea squirt (ascidian) to gain extra height. The sea anemone has feeding tentacles extended. (Credit: Simon Brockington, BAS); **(b)** A group of the Antarctic brachiopods *Liothyrella uva* adhering to the walls of a cave at a depth of 20 m. Two sea urchins (*Sterechinus neumayeri*) and a sea cucumber (*Cucumaria* spp.) can also be seen. Brachiopods are extremely slow growing and may live for up to 60 years. (Credit: Simon Brockington, BAS); **(c)** A group of sea urchins and soft corals in the shallows (12 m) in Marguerite Bay. (Credit: Simon Brockington, BAS); **(d)** Large red sea spider (Pycnogonida) walking over white hydroid. Antarctic sea spiders can be up to 50 cm across and have up to 12 legs. (Credit: Simon Brockington, BAS); **(e)** The octopus *Thaumeledone peninsulae*. (Credit: David Barnes, BAS)

(c)

(d)

(e)

suspension feeders, supported by food delivered in strong near-bottom currents. The other consists mainly of fauna living in the soft sediments and mobile epifauna living at the sediment surface in areas receiving high amounts of phytodetritus from the water column. However, these represent two extremes of a continuum, and intermediates also exist. Many benthic organisms living on the Antarctic continental shelf are characterised by gigantism, late maturity and longevity. While Southern Ocean deep-sea animals usually do not tend to display gigantism, we do not know to date whether late maturity and longevity apply for slope and deep-sea animals.

Unlike terrestrial or pelagic ecosystems, the benthos depends on energy input from the surface or the water column. Due to the variety of habitats available and the bottom topography, benthic communities are complex, very diverse and their trophic structures and the biogeochemical exchange between pelagic and benthic ecosystems are largely unknown for most areas. Two immediately striking features of this benthic diversity are that it is often extremely high – some groups such as pycnogonids being more diverse within the Antarctic/Southern Ocean than anywhere else globally – and it is very largely restricted to the Southern Ocean (endemic). The biomass and richness of some benthic communities is also often extremely high, in global terms second only to that of tropical coral reefs.

Over global geological timescales, plate tectonics, palaeoceanography and the resulting global climate changes (e.g. from greenhouse to icehouse) have had a strong impact on the Antarctic marine fauna and flora, including triggering the evolutionary radiation of many Antarctic benthic marine species. These have included events such as the progressive migration north of formerly cosmopolitan taxa established during the Jurassic and Cretaceous periods when Antarctica was still under greenhouse conditions, the creation of disjunct distributions due to plate tectonics (for instance, during the disintegration of Gondwana), the active migration of taxa in and out of the Southern Ocean and radiation events due to the emergence of new adaptive zones and habitats after the climate changed. Palaeoclimatic changes are thought to have driven latitudinal range shifts for many taxa, serving as 'taxonomic diversity pumps' into other regions.

As Antarctica is a large continent, the ocean and benthos surrounding it is actually not at such a high latitude as much of the Antarctic terrestrial biome, or the Arctic Ocean. The most southerly point in the ocean is found at 79°S (under a permanent ice shelf) in the Weddell Sea, still more than 2000 km from the South Pole. The continent's fringing sea ice, isolated by the ACC, generates cold and highly saline Antarctic Bottom Water, which is an important driver of global ocean circulation. The linked isolation of Antarctica's benthic faunas is due to temperature decreases in the Southern Ocean in the Cretaceous period, when many species became extinct while others displayed subsequent diversification. A few survivors, presumably with appropriate physiological pre-adaptations, then underwent a dramatic evolutionary radiation generating many key elements of the present-day benthic marine fauna. Evolutionary radiations have taken place in Antarctica in isolation over long periods of time, leading to typically high levels of species restricted to the area (60–90%). This is the case for sponges, bryozoans, polychaetes, pycnogonids, ascidians, anemones, peracarid crustaceans and some bivalves. The speciation processes in krill and notothenioid fish have received particular research attention, along with peracarid crustaceans and pycnogonids.

The Southern Ocean fish fauna has relatively low diversity, but it shows two striking radiations: the nototheniids on the continental shelf and the liparids on the continental slope. Therefore, the fish groups with the highest numbers of species today are the notothenioids, liparids and zoarcids, which between them account for 88% of species diversity; coincidentally, 88% of Antarctic fish species are also endemic. The level of dominance and the radiation of nototheniid fishes are unique and exhibit many parallels with species flocks of freshwater fishes in some ancient lakes. The notothenioid diversification was accompanied by considerable morphological and ecological changes. For instance, the most advanced adaptation to ice as a microhabitat for Antarctic fish has been observed for *Pagothenia borchgrevinki*, which has been reported to cling to the subsurface of the marginal ice shelf of the Drescher Inlet.

The isopod family Serolidae is likely to have originated between 90 and 55 million years ago. This estimate, based on linkages between geological processes and biogeographical patterns, roughly coincides with estimates for evolutionary radiations in some amphipod taxa. The inferred age of the last common ancestor of iphimediid amphipod species is 34.4 million years, approximately coinciding with the opening of the Drake Passage separating the Antarctic Peninsula from South America. The timeline of speciation, as well as their restriction to Antarctic waters, are both consistent with the view that the epimeriid and potentially also the iphimediid species evolved in the Southern Ocean when it was already cold and isolated from other fragments of Gondwana.

The Southern Ocean decapod fauna comprises only 98 benthic species in the entire region, with 92 species occurring on the continental shelves and around the sub-Antarctic islands and 6 species in the deep sea. The general global restriction of the Brachyura to shallow-water zones may underlie the absence of this group on Antarctic continental shelves, assuming successive eliminations of shallow water fauna by ice sheet expansion, scouring and anchor ice during glacial maxima. The striking Tertiary extinction and continued absence of the creeping decapods (e.g. crabs, lobsters) has been suggested as being due to physiological constraints of magnesium regulation in the cold water. The first recent records of anomuran and brachyuran larvae were obtained from the Bransfield Strait off the Antarctic

Peninsula, along with recent records of anomuran lithodid crabs in the deep sea and on the continental slope. These records raise the question of whether the return of the crabs to the Antarctic after their extinction in the Lower Miocene (-15 million years ago) might indicate a response to global change. However, it remains unclear when such recolonisation events actually took place and, in the case of the Bransfield Strait records, whether or not they involve any human activity in the transport process or, indeed, the species (which have been recorded only once) have actually become established. It is also plausible that periodic (re-)colonisation of the Antarctic continental shelf by species currently limited to the somewhat warmer adjacent slope and depths may be a more regular event associated with systematic incursions of warmer water masses onto the shelf. Currently, 11 species of lithodids have been reported south of the Antarctic Polar Front, and these have probably recolonised the Antarctic via the circumpolar deep sea.

The deep-sea benthos

Globally, the ocean depths rank amongst the least studied, visited or sampled ecosystems on the planet, despite the fact that they cover by far the largest geographic scale on Earth. A few decades ago it was suggested that marine biodiversity is higher in the deep-sea benthos than on the more familiar continental shelf or slope, and various hypotheses have been advanced to explain this. As it remains virtually impossible to obtain animals alive and undamaged from this region, we know virtually nothing in detail about the life histories, ecology or physiology of the deep-sea biota.

In contrast with the Antarctic continental shelf, which serves as an evolutionary laboratory due to its isolation and high degree of species endemism, the deep-sea regions of the different oceans are interconnected, and their fauna is not isolated. Vast areas surrounding the Antarctic continental shelf are deep sea. Compared to our knowledge of the benthos of the Southern Ocean shelf, knowledge of the deep-sea fauna remains scant, even after expeditions such as ANDEEP, focussing on specific regions, and the recent publication of more than 170 scientific papers.

Patterns of species diversity and benthic community structure differ between deep-sea habitats. The possible existence of latitudinal gradients in species diversity in marine ecosystems has been hotly debated, but in truth this issue currently remains unresolved, not least as vast areas of the deep sea, especially of the southern hemisphere, remain unexplored. Nevertheless, data for isopod and gastropod species richness challenge the hypothesis of latitudinal gradients in this region. High biodiversity has been reported for some macrofaunal and megafaunal taxa in the

Southern Ocean deep sea. Furthermore, in the Antarctic as has been seen elsewhere, it is becoming clear through the application of molecular biological techniques that some 'species' traditionally regarded as having circum-Antarctic distributions are in reality composed of several cryptic species (species which are morphologically alike and can only be discerned using molecular tools). The existence of such 'species flocks' in the deep-sea biota may be a more general feature than is currently appreciated. However, megafaunal diversity has been found to be distinctly lower in the Southern Ocean deep sea, especially below around 2300 m. The most important taxa here are the Porifera and Echinodermata, which are also known to be important in other deep-sea areas.

Through their production of Antarctic Deep Water, the Weddell and Ross seas may be important sources for taxa now living in the Atlantic or Pacific deep oceans, as the isothermal water masses surrounding the Antarctic continent provide an obvious conduit for the migration of shallow-water species into the deep sea and then transport further north. In this respect the Antarctic shelf might serve as a 'biodiversity pump', seeding diversification in the deep-sea fauna worldwide. Recent molecular data on the gene flow in bipolar species of single-celled Foraminifera support this hypothesis, as do data on pressure tolerance of early larval stages of the Antarctic sea urchin *Sterechinus neumayeri* which can persist to at least 2000 m depth.

Given the continuing lack of comprehensive survey and biogeographic data, even now it remains impossible to assess properly levels of endemism in the Southern Ocean deep sea. There do appear to be systematic differences between biological groups related to their reproductive mode. For instance, brooding isopods, which should have reduced possibility for gene flow, occur more locally with a high number of species and many rare, apparently 'endemic', species in the deep sea (−85%). Levels of species endemism for both shelled gastropods and bivalves are approximately 75%, while other taxa with free-living larvae, such as polychaetes, have apparently much wider zoogeographic distributions and even lower degrees of endemism in the Southern Ocean deep sea than on the shelf.

While the ANDEEP project has revealed patterns of biodiversity within different faunal groups and documented that these can vary significantly, we still know very little about the ecology and role of deep-sea fauna in trophodynamic coupling and nutrient cycling in oceanic ecosystems. To fill this knowledge gap, a successor to the ANDEEP project, the ANDEEP-SYSTCO (SYSTem COupling) has been started within the framework of the International Polar Year (IPY). SYSTCO aims to investigate the functional biodiversity and the ecology of dominant abyssal species and examine the trophic structure and functioning of abyssal communities of the Atlantic sector of the Southern Ocean, focussing on the role and feeding of the abundant key organisms. Investigations of the Southern Ocean deep-sea food web

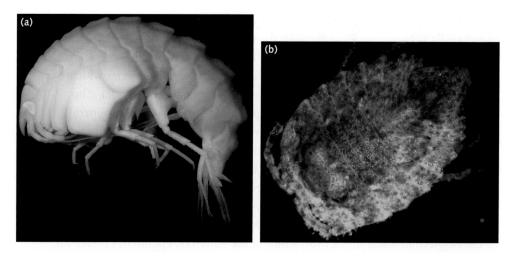

Figure 6.16 (a) The amphipod *Epimeria angelikae* Lörz & Linse 2011. (Credit: Anne-Nina Lörz & Katrin Linse); (b) A species of isopod *Frontoserolis* (Serolidae). (Credit: Wiebke Brökeland)

done using fatty-acid patterns of peracarid crustaceans as well as stable isotope ratios have revealed that some species feed on a wide variety of different food items, especially foraminiferans, a potential reason why some taxa of Isopoda or Amphipoda might be so successful in the Southern Ocean and have diversified.

Recruitment and dispersal

Most marine invertebrates disperse by means of a planktotrophic (feeding in the water column from an early stage of development) or lecithotrophic (early development depends on egg yolk stores) larva. The accepted paradigm is that the former permits wider dispersal, although this has been challenged as at least some lecithotrophic larvae have potentially longer larval life in the plankton than planktotrophic species. Any such dispersal adaptations would be beneficial for dispersal in the generally nutrient-poor deep sea, but little is known of reproduction in these deep-sea species when compared to the world ocean.

Antarctic shallow water environments are amongst the most disturbed in the world through ice action, and are characterised by a high proportion of species with protected and pelagic lecithotrophic development – such an environment obviously favours organisms with high dispersal capabilities. In addition there is a predominance of broadcasters (e.g. *Sterechinus neumayeri* and *Odonaster validus*) at locations where disturbance is common, while brooders (e.g. *Abatus agassizii*) only occur at shallower depths in the least disturbed locations.

Figure 6.17 Sea urchins caught with the sediment profile camera in the central Powell Basin at 1955 m. Urchins of 5 cm in diameter sit on a silty clay with some coarse-sand sediment, which was disturbed when the camera entered the sediment. (Credit: Robert Diaz)

As early as 1936, Gunnar Thorson proposed that polar environmental conditions, whose influence has been felt since the Cretaceous, had a particular influence on sensitive early life-history stages and favoured brooding. In general, biogeographic patterns of taxa may be linked to larval ecology. Groups such as the isopods, ostracods and nematodes have poor dispersal capabilities and hence reduced gene flow, making restricted species distributions more plausible, albeit difficult to prove. In contrast, other groups such as polychaete worms appear to be able to cross the barrier between the Southern Ocean and adjacent oceans and have a much wider zoogeographic and bathymetric distribution. Brooding is thought to be much less common amongst polychaetes, making dispersal via their free-swimming trochophore larvae more common. Polychaetes may also have particularly high physiological flexibility in coping with temperature and pressure changes.

Cephalopods and other molluscs likewise have a wider zoogeographic distribution and a lower degree of endemism.

Wide bathymetric ranges (eurybathy)

During the Cenozoic, and particularly during the Miocene and Pleistocene (23–5 million years ago), considerable extensions of the Antarctic ice sheet across the continental shelf have clearly had drastic impacts on the shelf biota. There is geomorphological evidence of ice scouring to depths of at least 1000 m in places, and it is generally accepted that the ice shelves extended uniformly to the edge of the continental shelf around the Antarctic Peninsula and also much of East Antarctica. Although it has been suggested that this has led to periodic complete eradication of the shelf fauna it is very likely that this has not happened. However, the wider bathymetric distribution of Antarctic invertebrates when compared to other seas does suggest that many continental slope species found shelter in deeper waters off the shelf during these glaciation events. The historical changes in ice-shelf extent probably led to adaptive radiations and speciation processes on the Antarctic continental shelf and slope, again in effect generating a 'biodiversity pump'. While we cannot quantify directly the extent to which species have migrated up and down the Antarctic continental shelf and slope, these processes are likely to have selected for the wide bathymetric ranges (eurybathy) that are characteristic of many Southern Ocean species today. The boundary between the shelf /slope and true deep-sea marine faunas lies between 1500 and 2500 m depth in the Weddell Sea. At these depths, groups such as Foraminifera show close links with deep-water faunas of other oceanic regions, perhaps through the influence of the thermohaline circulation. However, cryptic speciation has also been documented for Foraminifera, suggesting a relationship between populations from widely separated geographic regions of the Southern Ocean (e.g. Weddell and Ross Seas), while other species are adapted to a wide bathymetric range.

Macroalgae

The Antarctic macroalgal flora and its relationship with those of surrounding continents is well characterised, and provides an important structural component of nearshore marine ecosystems particularly of the sub-Antarctic islands. The Desmarestiaceae, which originated in the southern hemisphere, occur worldwide. Bipolar species developed in either of the two hemispheres and crossed the equator during the Last Glacial Maximum. Polar macroalgae are adapted to weak light conditions but, as with plants on land, in order to obtain the light required for photosynthesis, exposure to UV radiation is inevitable. Brown, green and red algae

Figure 6.18 Kelp beds around the coast of Iles Crozet, which are important spawning grounds for many Antarctic fish. (Credit: D. W. H. Walton)

are most sensitive to UV in the littoral zone, but most species also occur in the sublittoral. It seems that UV sensitivity, along with ice scouring, is an important factor influencing the zonation of these macroalgae along the coast. Dispersing cells are also adapted to function in low light conditions, with photosynthesis further depressed by exposure to UV radiation. Even with the possession of chemical defence mechanisms against invertebrate grazers, macroalgae are centrally important primary producers – given their often remarkable abundance they are likely to play a significant role in mediating energy and nutrient transfer in the nearshore Antarctic environment. For many benthic animals, e.g. echinoderms, molluscs, peracarid crustaceans and fish, macroalgae are important not only as a food source, also providing shelter and space to hide for the offspring.

Terrestrial ecosystems

At first sight, the terrestrial ecosystems of Antarctica appear to provide a direct and striking contrast to those of the surrounding benthic and pelagic ocean realms. While the latter are extensive, diverse, include great biomass and have an undeniably long evolutionary history, habitats on land are sparse and visually barren, biota are miniscule, their diversity is low and, at least until recently, any historical signal they provide was thought to be insignificant. Today's Antarctic terrestrial ecosystems are isolated from other continents and from each other. Only about 0.34% of the continental area is not permanently covered by snow or ice, including nunataks, cliffs and seasonally snow and ice-free areas. Most terrestrial habitats are found in coastal regions, primarily the Antarctic Peninsula, along with a series of isolated oases along the coastline of continental Antarctica, and in major mountain ranges such as the Transantarctic Mountains. Most of these terrestrial ecosystems

Figure 6.19 Some sub-Antarctic terrestrial habitats: **(a)** the dwarf shrub *Acaena magellanica* dominates large areas on South Georgia, its hooked seeds allowing it to be spread by birds (Credit: Peter Convey, BAS); **(b)** Tussac grass *Parodiochloa flabellata* forms a coastal fringe on South Georgia, and is a major nesting habitat for burrowing petrels (Credit: D. W. H. Walton, BAS); **(c)** Mats of the green alga *Prasiola crispa*, which is the only species able to grow on the nutrient rich outflow from a penguin rookery (Credit: Roger Worland, BAS);

(d) The Kerguelen cabbage *Pringlea antiscorbutica* is a long-lived species and plays a major role in the structure of the herb associations on Iles Crozet (Credit: D. W H Walton, BAS); **(e)** *Poa cookii* grassland on Iles Crozet. (Credit: D. W. H. Walton, BAS)

are effectively islands, whether surrounded by ice or sea. One striking exception to this generalisation is provided by the 'Dry Valleys' of Victoria Land, where ablation in the incredibly dry air outweighs precipitation, leaving a frigid desert area tens of thousands of square kilometres in extent.

Antarctica is an ice-bound continental mass surrounded and isolated by a large extent of cold ocean. This contrasts with the Arctic, where large and ice-free continental masses surround a polar ocean. This simple geographical contrast underlies fundamental climatic and biological differences between the two. At any given latitude, the climate experienced on land in Antarctica is considerably more harsh, while the extreme isolation from possible refugia and other southern hemisphere sources of colonists, combined with large-scale (but not complete) obliteration during Pleistocene and previous glacial maxima, have minimised colonisation and led to very low levels of diversity for groups present.

Antarctica was a central component of the supercontinent Gondwana. As Gondwana fragmented, its last continental links were with Australia and South America, and these were lost 30–35 million years ago. Fossil evidence from that time indicates that Antarctica had a cool temperate fauna and flora, and that elements of this persisted as a tundra-like assemblage until at least the mid Miocene, 12–14 million years ago. The development of circumpolar ocean currents and atmospheric circulation patterns had physical as well as biological consequences, accelerating the continent's gradual cooling and, eventually, glaciation. The extent of continental cover by ice sheets and glaciers has varied widely over this period, although Pleistocene glaciation mirrored that of the Arctic. The majority of time over at least the last 800 000 years has been spent in the glacial phase of glacial–interglacial cycles, with periods of considerably thicker ice sheets than now, and with vast ice shelves extending out to the edge of the continental shelf. Periods of ice maxima are often thought to have led to complete eradication of terrestrial biota (not least, simply through loss of habitat), followed by waves of recolonisation. However, recent work across the continent has identified many examples of biota whose presence is most easily explained as persisting through the ice ages in refugia rather than as colonists, and reconciling these observations with glacial reconstructions is a pressing challenge.

At a broad scale, Antarctica's terrestrial ecosystems and biology are treated as three components, whose ecosystems are distinctly different. However, while convenient and useful, this remains an oversimplification and is currently an active area of scientific advance.

The richest terrestrial ecosystems are found in the sub-Antarctic, which consists of a set of isolated Southern Ocean islands and archipelagos, including South Georgia, the Prince Edward Islands, Îles Kerguelen and Crozet, Heard and Macdonald Islands, and Macquarie Island. The majority of these lie either close to or north of the oceanic Polar Frontal Zone. Further island groups are sometimes included under the term 'sub-Antarctic', having some similarities in terms of biota present and climatic regimes (Juan Fernandez, Falkland Islands, Gough Island, Îles Amsterdam and St. Paul, New Zealand's outlying groups of Snares, Campbell, Bounty, Auckland and Chatham) although they are more correctly described as southern cool temperate or ocean temperate. While they do share many biological similarities with the true sub-Antarctic islands, they also host faunal and floral groups not otherwise represented in the sub-Antarctic. All sub-Antarctic island climates are strongly oceanic with mean air temperatures low and positive year round, high precipitation and, except for South Georgia, they do not come under the influence of seasonal pack or fast ice.

The maritime Antarctic consists of a western coastal strip along the Antarctic Peninsula, along with the Scotia Arc's South Shetland, South Orkney and South Sandwich archipelagos, and the isolated Bouvetøya and Peter I Øya. Conditions are more extreme, with mean temperatures slightly above 0°C for 1–4 months of the year, although like the sub-Antarctic they are buffered from extremes by the surrounding ocean, which varies annually between c. −2 and +1°C at shallow depths. Precipitation is, again, generally high. This region includes some unique geothermal terrestrial ecosystems associated with volcanic activity on the South Sandwich and South Shetland archipelagos and Bouvetøya.

The continental Antarctic is the third commonly recognised biological region. Ostensibly encompassing a much greater area, it conventionally includes all East (or Greater) Antarctica, the Balleny Islands, and the eastern side of the Antarctic Peninsula. However, other than the Victoria Land Dry Valleys, the total area of exposed terrestrial habitat is extremely limited. Conditions are again more extreme, with positive mean air temperatures recorded for less than one month each year in coastal locations and never inland. Away from the coast, air temperatures rarely become positive even for short periods in the diurnal cycle in summer, although absorption of solar energy by even small dark surfaces means that conditions at the micro scale (appropriate to the biota) are less extreme in summer than the standard meteorological records suggest.

Terrestrial biota

Other than on the sub-Antarctic islands, most terrestrial habitats experience seasonal snow or ice cover, giving protection from temperature extremes and wind abrasion. The duration of snow- or ice-free periods in summer varies widely, from periods of days or weeks up to several months. Sub-Antarctic islands may only have intermittent snow cover, and then often only at altitude. On these islands microhabitat temperatures in soil or vegetation, with or without snow cover,

Figure 6.20 Vegetation around a geothermal vent in the South Sandwich Islands. (Credit: Peter Convey, BAS)

typically remain sufficient to allow year-round biological activity. In contrast, along the Antarctic Peninsula and continent, biological activity is arrested by low winter temperatures.

Antarctic soils are typically poorly developed, with low organic content, and are heavily influenced by disturbance and mixing during frequent cycles of freezing and thawing (cryoturbation). Permafrost is not present in the sub-Antarctic at least up to several hundred metres altitude but generally underlies ice-free ground across the continent. Brown soils are reasonably well developed in the sub-Antarctic, but are found only with larger stands of higher (flowering) plants in the maritime Antarctic, and not at all on the main body of the continent. The presence of well-developed moss communities has led to deep peat deposits in the sub-Antarctic, dated to soon after the end of the Pleistocene glaciation. Such deposits are unusual and younger in the maritime zone (up to 5–6000 years old), and are not found in the continental Antarctic.

The apparent abundance and biomass of birds and seals on the sub-Antarctic islands and parts of the continental coastline is deceptive, as these are a component

Figure 6.21 Periglacial sorted circles, Davis Valley, produced by ice wedging in the soil forcing smaller rocks into the centre. (Credit: Peter Convey, BAS)

of the marine ecosystem. Nevertheless, they have a considerable impact on terrestrial ecosystems on the local scale, through their input of nutrients through manuring, and the physical impact of trampling. More remote impacts are also detectable through aerosol transfer of nutrients. In contrast, true terrestrial vertebrates are extremely limited, including a single endemic passerine (insectivorous) bird and three freshwater ducks in the sub-Antarctic and two scavenging sheathbills, one of which is also found in the maritime Antarctic. Terrestrial mammals, reptiles and amphibians are absent.

In the absence of vertebrates, invertebrates dominate Antarctic terrestrial ecosystems. Even in the more benign sub-Antarctic many groups are not represented, and the most diverse insect groups are Diptera (flies) and Coleoptera (beetles). Molluscs and annelids are present along with the micro-arthropod groups of Acari (mites) and Collembola (springtails) and micro-invertebrates (nematodes, tardigrades, rotifers). In concert with conditions becoming more extreme, maritime Antarctic faunas include few higher insects, with only two chironomid midges present. Mites and springtails can be abundant, with high densities comparable with many temperate ecosystems, and micro-invertebrates are also numerically abundant. In all these groups, species diversity is low. Food webs are characteristically simple, with most energy flow through the decomposition cycle and negligible impacts of predation or true herbivory. The continental Antarctic provides an even simpler fauna, with no insects, and much more limited presence of micro-arthropods. The simplest faunal ecosystems yet known on the planet are found here, where even nematodes are absent.

The Antarctic flora is generally well described, and shares several features with the fauna. The sub-Antarctic is again richest, and shows relationships with the floras of the relevant nearest continents or landmasses. The strictly defined sub-Antarctic is characterised by the absence of trees and shrubs. Although often described as comparable with Arctic tundra, the sub-Antarctic differs in important features, in particular in not being influenced by permafrost and in the absence of woody plants. A particular feature of sub-Antarctic vegetation is the presence of large 'megaherbs' and tall grasses, whose evolution is thought to have been encouraged by the complete lack of vertebrate grazers.

Vegetation communities of the maritime Antarctic, along with higher altitude areas of the sub-Antarctic and much more limited coastal areas of East Antarctica, host cryptogamic communities of carpet- and turf-forming mosses. Extensive areas of vegetation are only found over a narrow altitudinal range and near the coast. Only two flowering plants are present on the Antarctic continent, both limited to the Antarctic Peninsula, and both also found in the Andes and Tierra del Fuego of South America – these are the Antarctic hairgrass, *Deschampsia antarctica*, and Antarctic pearlwort, *Colobanthus quitensis*. As conditions become more extreme, particularly moving into the continental Antarctic, closed moss turves are replaced by open fellfield moss and lichen communities, and mosses gradually decrease in abundance and diversity.

While there is implicit recognition within the descriptions above that microbial groups must make an important contribution within Antarctic terrestrial ecosystems, along with a more general recognition that microbial activity is fundamental to ecosystem processes, the microbial elements of Antarctic ecosystems have received surprisingly limited attention to date. Microbes are central to the primary colonisation and stabilisation of mineral soils, subsequently permitting colonisation and succession by other biota. Microbes also are the only living components of some of the most physically extreme ecosystems on Earth that are found in Antarctica. These include habitats described as cryophilic (between ice crystals), chasmoendolithic (within tiny cracks that are open to the rock surface) and cryptoendolithic (within microscopic cavities of the rock matrix).

Figure 6.22 Some of the Antarctic terrestrial invertebrates found in the soil:
(a) Nematode (Credit: BAS); **(b)** an aggregation of springtails (Collembola) floating on a water film (Credit: BAS); **(c)** Antarctic oribatid mites *Alaskozetes antarcticus*, a group of adult (black) and juvenile along with predatory mesostigmatid mites *Gamasellus racovitzai* (orange) (Credit: BAS); **(d)** mating pair of flightless Antarctic midge *Belgica antarctica*, Antarctica's largest land animal (Credit: R. E. Lee).

Figure 6.23 Some Antarctic terrestrial vegetation types: **(a)** The two native flowering plants found in Antarctica – *Deschampsia antarctica* and *Colobanthus quitensis* – photographed in the South Orkney Islands (Credit: Peter Convey, BAS); **(b)** *Polytrichum* moss banks on the South Sandwich Islands (Credit: Peter Convey, BAS); **(c)** Cyanobacterial mat in a small pool on Adelaide Island (Credit: Peter Convey, BAS); **(d)** Crustose lichens, the most diverse and widespread plants in Antarctica, on coastal rocks of Léonie Island, Marguerite Bay (Credit: Peter Bucktrout, BAS).

(d)

Figure 6.24 Cryptoendolithic community living within Beacon sandstone in the Dry Valleys. The green layer towards the top is of algae and cyanobacteria, the white layer in the centre is fungus and the black layer at the bottom is the two combined as a lichen. These communities may grow for only a few hours each year when there is liquid water and warmer temperatures and thus may be many thousands of years old (Credit: Chris Gilbert, BAS).

Freshwaters

Freshwaters are poorly represented in the Antarctic relative to other continents. There are few true rivers, and these are short – the longest being the few kilometres of the Onyx River in the Dry Valleys – although there are many ephemeral melt streams during the short Antarctic summer. Freshwater lakes have been subject to detailed study on some sub-Antarctic islands, in parts of the maritime Antarctic (South Orkney and South Shetland Islands), along the continental coastline (Larsemann and Davis areas, Syowa) and in Victoria Land. The latter region includes some particularly unusual lakes within the Dry Valleys, ranging from freshwater to hypersaline. As with terrestrial ecosystems, those of freshwaters have a very simple structure. There is again an absence of vertebrates (fish, amphibians). Higher trophic levels are again very limited – for instance there is only a single predatory diving beetle, and that is limited to sub-Antarctic South Georgia, a single predatory copepod in the maritime Antarctic, and no predatory Crustacea in the whole of the continental Antarctic. Microbial foodwebs assume an importance

Figure 6.25 Lake Fryxell, Taylor Valley, Victoria Land. This is one of several Dry Valley lakes that melts sufficiently during summer to have open water around the edge (Credit: Brien Barnett, NSF).

in Antarctic freshwaters that is not generally seen at lower latitudes. 'Top down' grazing control is very limited, and the 'microbial loop' predominates. This consists of energy and nutrients flowing through a food web composed of microscopic phytoplankton, bacteria and protozoans.

As with the nearshore and littoral marine environments, ice plays a very large role in the biology of Antarctic lakes. In the maritime and continental zones, and the colder sub-Antarctic islands, lakes are seasonally or permanently covered by ice. This has diverse consequences, affecting, for instance, light transmission, water column stability, thermal and chemical stratification and oxygen availability, as well as the direct physical damage from ice nucleation in contact with biological material, and damage from moving ice. While many lakes are obviously recent in origin, there is evidence that some have persisted through at least one full glacial cycle, while some in the Dry Valleys are thought to be tens or even hundreds of thousands of years old. In recent years the existence of many subglacial lakes deep under the Antarctic continental ice sheets has come to prominence. The largest of these is Lake Vostok, 200 km long and 500 m deep, lying below at least 3.6 km thickness of the continental icecap. Initially thought to have been completely isolated from the modern world since the formation of their overlying ice sheets, it is now known that many subglacial lakes are linked by under-ice water flows, but they still harbour the intriguing possibility of containing microbial communities that have evolved in isolation for hundreds of thousands of years.

Biological history and colonisation processes

The Antarctic terrestrial biota has obvious linkages with the other southern continents, particularly evident in the flora. This, along with the perceived wiping out of terrestrial habitats as ice sheets expanded at glacial maxima, has led to a view that today's biota is largely a result of recent colonisation as post-Pleistocene deglaciation progressed. Despite the continent's geographical isolation and extreme conditions, colonisation processes plainly continue to operate, albeit at very low frequency. Some Antarctic biological groups appear to possess features that are supremely well adapted to facilitate long-distance dispersal, for instance, the spores and vegetative propagules of mosses and lichens, the resting tuns of tardigrades and the many well-expressed ecophysiological adaptations to cold and desiccation stresses. There are well-documented instances of plant and microbial groups found in Antarctica (both widespread and as single records) with wider bipolar or temperate/tropical montane distributions.

Amongst many of the groups of biota that dominate Antarctic terrestrial ecosystems (mites, springtails, tardigrades, nematodes, lichens, but not mosses) there are many examples of high levels of endemism, both to the Antarctic in general, and to specific subregions within Antarctica. These are now used to support a hypothesis of prolonged regional presence ('vicariance') rather than recent colonisation. Molecular biological techniques have given a new approach to estimating divergence dates between related organisms. These have demonstrated radiation processes over millions of years in the springtails of the Transantarctic Mountains and Victoria Land, and even older separation events in the Antarctic Peninsula and Scotia Arc, with divergence of closely related species of endemic flies possibly coincident with the tectonic separation of the Antarctic Peninsula from South America on timescales of 30–40 million years or more.

While the strength of evidence supporting the generality of vicariance as underlying a large proportion of the present day terrestrial biota of Antarctica is rapidly increasing, there has been less progress in identifying specific refugial locations. Indeed, this may never be possible, as it is a feature of repeated glacial and ice sheet advance and retreat cycles that the small-scale detail of ice boundaries is very variable, while successive advances obliterate evidence of previous boundaries. It is undoubtedly the case that, at glacial maxima, far less exposed terrestrial ground was available than is the case even now. This inevitably means that, with each cycle of glacial retreat comes the opportunity for 'local' recolonisation from refugia – this, for instance, must underlie the contemporary

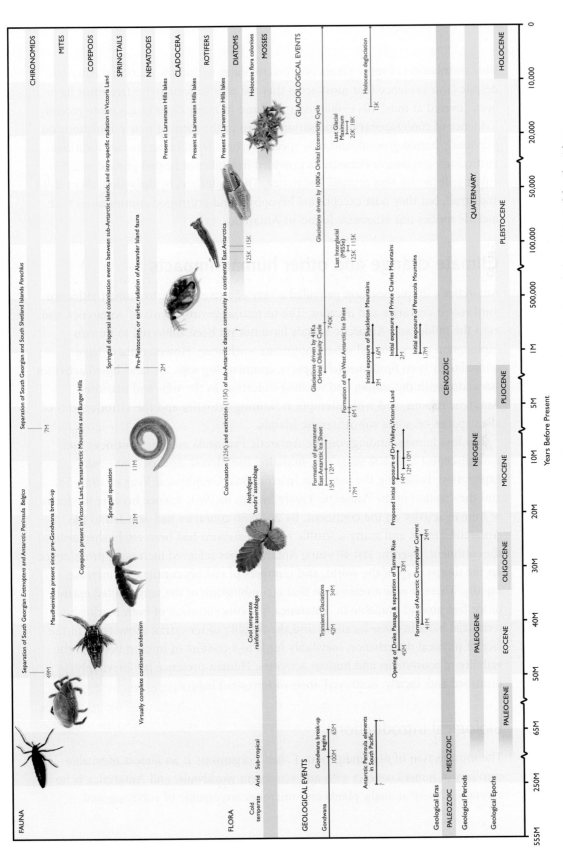

Figure 6.26 Biological colonisations and extinctions in Antarctica inferred from biogeographic, molecular, phylogenetic and fossil evidence. (Credit: Peter Convey, BAS and Blackwell-Wiley)

wide distribution of most plants and invertebrates within the maritime Antarctic, despite clear evidence that most areas that are now seasonally ice free must have been covered at most 10–15 000 years ago, and in many cases much more recently. Evidence of contemporary colonisation is seen in the form of new populations on previously barren ground, and new species records from better-known locations. Particularly impressive evidence is provided in studies of heated ground at a few volcanically active sites around Antarctica. These sites are, by definition, short-lived and small, but they host exceptional bryophyte and arthropod communities that include species not otherwise found in Antarctica.

Climate change and other human impacts

Antarctica's simple isolation provided a very effective barrier to human influence until recent centuries and decades. The terrestrial environments of Antarctica and even the milder sub-Antarctic islands have not yet been subjected to human exploitation as seen on all other continents worldwide. However, there have undoubtedly been fundamental impacts, commencing with the land-based activities associated with the sealing and whaling industries in the sub- and maritime Antarctic regions, and with attempts at farming activities and the introductions of alien species on some sub-Antarctic islands.

The first human landings on the Antarctic Peninsula and the continent itself took place just over 100 years ago, initiating the 'Heroic Age' of continental exploration. Following the post-war International Geophysical Year of 1957–58, and the creation of the Antarctic Treaty System in 1961, science became the focus of human activity on the continent. By 2012, 50 countries had acceded to the Antarctic Treaty, and many scientific research stations had been established around the continent. Over the last 40 years, Antarctica has achieved increasing prominence on the tourist map of the world, and numbers of visitors continue to increase rapidly. There is now a realisation that a combination of the very limited extent of ice-free ground available in Antarctica, the concentration of both marine and terrestrial biota in these locations and the fragility of terrestrial ecosystems in the face of physical disturbance, inevitably leads to a conflict of interest between the viability of ecosystems and human activities. Human presence has inevitably disturbed and, locally, destroyed areas of terrestrial habitat.

Biological introductions

The introduction of non-indigenous ('alien') organisms is an almost inevitable corollary of human contact with any ecosystem worldwide, and Antarctica is no exception. Many animals, plants and microbes are capable of surviving and establishing if they are accidentally or deliberately transferred to Antarctica. This is abundantly clear in sub-Antarctic terrestrial ecosystems, which have had the longest record of human influence and have the least extreme environmental conditions. A range of vertebrates are present through both accidental (rodents) and deliberate (farm animals, pets) introduction to most major islands. Although there have been some control and eradication attempts, for example of cats from Marion and Macquarie Islands and rabbits and rats from islands in the Kerguelen archipelago, the consequences of these introductions may now be irreversible whilst, even if possible in theory, the practicalities and costs of achieving effective control measures for most species may be insurmountable. The impacts of some (particularly predators and grazing mammals) are dramatic. There are also a wide diversity of introduced plants on the sub-Antarctic islands, some of which may prove capable of spreading further south as summer conditions warm.

In the marine environment, while there are currently very few confirmed records of introductions, our practical ability even to detect them is minimal, given the limited spatial extent of survey data and equipment. In an analogous fashion on land, and almost inevitably, introductions of less visibly obvious organisms (invertebrates, lower plants, microbes) have received less or no attention. However, it is clear that the vast majority of introductions overall have been accidental. While the majority have existed only in concert with human occupation or activity, some of those that have become established have gone on to become invasive, with clear and harmful impacts on native ecosystems especially on the sub-Antarctic islands. There is evidence of competitive displacement and local extinction of native species, and in this context the introduction of new trophic levels to ecosystems where they were previously unrepresented, such as predatory carabid beetles to two sub-Antarctic islands, is particularly significant.

Climate change and other human impacts

Antarctica has been a focus of biological studies aimed at understanding the consequences of global climate change. Parts of the continent, in particular the Antarctic Peninsula and Scotia Arc islands, have experienced the most rapid temperature increases seen worldwide over the last 50 years or so. The polar regions are also predicted to see the greatest and most rapid warming in future. Coupled with this, the relative simplicity of their terrestrial ecosystems is seen as a tool with which to identify more easily the ways in which ecosystems generally may be vulnerable to or respond to processes of environmental change. While there is an almost inevitable media focus on climate warming, climate change is a far more complex process. For terrestrial biota, a second fundamentally important change lies in precipitation patterns and form (snow vs. rain) as this alters water input to

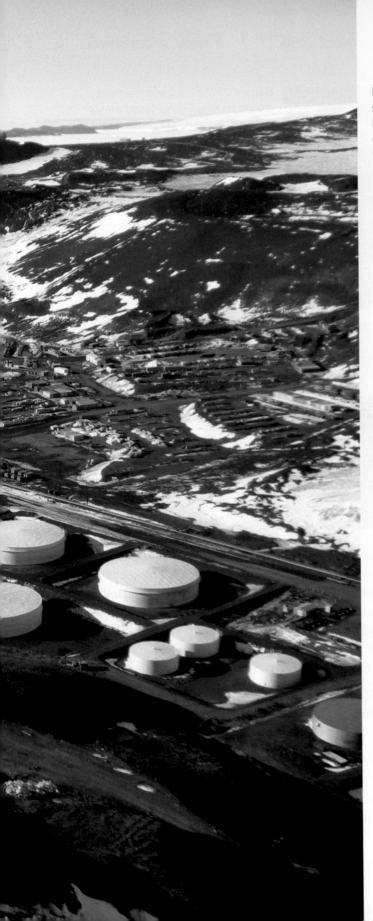

Figure 6.27 McMurdo station, Ross Island, showing widespread impacts of station buildings and facilities. (Credit: Peter Convey, BAS)

Figure 6.28 Vehicle tracks over wet ground on King George Island damage the moss patches and may persist for many years. (Credit: Peter Convey, BAS)

Figure 6.29 Grazing exclosure on South Georgia showing the extent of regeneration once the reindeer are excluded, although the species diversity is often permanently changed. (Credit: Peter Convey, BAS)

terrestrial habitats, as do other local changes such as the exhaustion of a snow bank. In the sea, changes in patterns of ice cover, formation and loss, all have potentially wide ranging consequences for biota. Finally, the impact of ozone hole-related increases in harmful short-wave UV-B radiation is often mistakenly confused with global warming processes. This anthropogenic influence on Antarctic ecosystems, catalysed inadvertently by the release of aerosol propellants and refrigerant gases into the atmosphere, has existed since the early 1980s.

Biological consequences of environmental change

At high latitude sites with very restricted thermal energy budgets, or in thermally very stable environments, the relative importance of a small temperature increment will be greater. On land, the main consequences of warming will be to increase the length of the growing season, giving increased growth, reproduction and population sizes. In contrast, in the sea, it has been suggested that major components of marine

Figure 6.30 A diversity of introduced plant species growing amongst the buildings at Grytviken whaling station on South Georgia. Whilst many of the plants flower every year only some species are able to produce viable seed. (Credit: Peter Convey, BAS)

ecosystems may rapidly exceed the thermal limits within which they continue to be viable. Other changes could also be harmful for Antarctic biota. Decreases in water availability on land increase the risk of local extinction and change in ecosystem structure. Similarly, if exposure to UV-B radiation passes biological thresholds, some ecological processes may be compromised, such as the central microbial role in soil stabilisation on land, and algal primary production in the sea. Acidification might become one of the largest problems for the Southern Ocean ecosystem, in both the pelagic and benthic realm. In practice, responses to all these changes are likely to be complex and hard to quantify, involving changes in patterns of investment in ecophysiological, biochemical and more general elements of life history.

As a consequence of their highly stressful and variable environment, many indigenous Antarctic terrestrial biota have inherent ecophysiological flexibility and can absorb and benefit from some of the effects of predicted levels of change, at least in the absence of newly colonising competitors. However, it is also clear that any amelioration is likely to reduce the constraints or barriers that limit colonisation and establishment of new non-indigenous species. Ecophysiological research on organisms of both land and sea is proving to be central to advancing understanding of the mechanisms of climate change impacts and responses, of global relevance to biota. It is also very simplistic to look at the individual species level alone, and improving knowledge of how these responses may be integrated to higher levels within the ecosystem is now an urgent priority. Antarctic research, drawing on the advantages of relatively simple and well-understood ecosystems, can again be central to achieving this.

As is already being seen along the Antarctic Peninsula, regional climate change (warming) is leading to widespread ice retreat, increasing the area of available terrestrial habitat, and season length. New ground is colonised rapidly by native biota, and existing terrestrial ecosystems also increase in biomass and development. With or without further human assistance, this will result in increased biodiversity and community development. The almost inevitable colonisation and establishment of new non-indigenous species will, in turn, generate greater trophic complexity and involve higher trophic levels. The resistance of indigenous communities to these processes is unknown.

The global importance of Antarctica

Antarctica, and in particular the many and varied facets of its biology, has always acted as a magnet for researchers, natural historians, explorers and the armchair enthusiast. Even today, in an age of easy global travel and instant information, we are still fascinated by its extreme challenges. In research terms, understanding of 'life at the extremes' is central to any attempt to describe limits and function of life on Earth, at levels from the gene, through genomic expression, biochemistry, physiology and upwards through life history to community and ecosystem function and macroecology. Antarctic ecosystems are also central to public fascination, and, in particular, a burgeoning tourism industry. Specific sub-Antarctic islands, under national jurisdictions, have already been declared as World Heritage Sites, with the concomitant responsibilities this places on nations towards their preservation for the benefit of present and future generations. Similarly, within and around the continent itself, many 'Antarctic Specially Protected Areas' have been declared under the auspices of the international Antarctic Treaty System that regulates activities on the continent.

Antarctica today has never been of greater value to the planet. This continent's near pristine environment and lack of confounding influences make it our best barometer of the state of the global climate – a 'canary in the coal mine' warning of the consequences of human actions for the planet. Its terrestrial and marine ecosystems are central in the conservation of global biodiversity, both in the context of the well-known and highly visible 'charismatic megafauna', and of the much less appreciated but equally important ecosystem elements that speak of the evolutionary history of the continent and Southern Ocean. Antarctic biodiversity is also starting to receive attention in the newer industrial concepts of bioprospecting and applications within biotechnology. Despite the spectacular and sorry history of overexploitation and mismanagement of marine mammal and fishery resources, the Southern Ocean of today still contains one of the last major and largely unexploited sources of marine protein available on the planet, as well as one of the few ecosystem-based and internationally agreed systems of management of marine resource exploitation. While we may feel uncomfortable with this 'opening up' of Antarctica, areas of exploitation such as these now appear inevitable, and the next few years and decades will provide one of the last chances available to humankind to demonstrate an ability to do so 'sustainably', and without destruction of the very resource that we seek to protect.

7

Space science research from Antarctica

LOUIS J. LANZEROTTI AND ALLAN T. WEATHERWAX

> The eastern sky was massed with swaying auroral light, the most vivid and beautiful display I have ever seen – fold on fold the arches and curtains of vibrating luminosity rose and spread across the sky, to slowly fade and yet again spring to glowing life. **R. F. Scott, 22 June 1911**

Introduction

At the ends of the Earth it is not only the land and the seas that are unusual. Even the upper atmosphere is different, providing remarkable insights into the way in which the planet interacts with the Sun and evidence for our wandering magnetic fields. The most striking visible evidence of these phenomena are the lights in the winter sky – the Aurora Australis – produced by the interaction of solar emissions with Earth's environment, and first seen by European explorers in 1773 by Captain James Cook's expedition. Of course, this southern equivalent of the 'northern lights' was observed on occasion by native peoples of the southern regions of New Zealand and Australia. To the Maoris, the aurora was known as Tahu-Nui-A-Rangi, a red glowing of the sky resulting from reflections of great fires. Similarly, the aboriginal people in Australia viewed the aurora as the feast fires of the Oola-pikka, ghostly spirits who spoke to the elders through these auroral flames. Modern explanations, albeit less fanciful, are certainly just as colourful, and centre on complex electromagnetic interactions between the Sun and Earth.

Whilst the first description of the effects of the Earth's magnetic field on a freely suspended needle are attributed to the Chinese in the twelfth century it was not until the fifteenth and sixteenth centuries that seafarers began to appreciate the importance of the magnetic poles. Increasing scientific interest in magnetism, in due

Antarctica: Global Science from a Frozen Continent, ed. David W. H. Walton. Published by Cambridge University Press. © Cambridge University Press 2013.

Figure 7.1 British naval officer and polar explorer, Sir James Clark Ross (b. 1800). A pioneer in the study of Earth's magnetic field, he is shown here with a dip circle, an instrument used to measure the angle between the Earth's magnetic field and the horizon. Although he discovered the North Magnetic Pole, the southern pole remained elusive. That discovery went to the Australian geologist Sir Douglas Mawson in 1909.

course, led to organised magnetic observations. Beginning in the first half of the nineteenth century and continuing ever since, observations of Earth's magnetic field, including the three vector – directional – components, were begun on a more-or-less regular basis in Europe (especially in Germany under the strong guidance of Carl Frederick Gauss and Wilhelm Weber), and in Britain by Edward Sabine. These observations persuaded a number of eminent natural scientists to promote, in the early nineteenth century, global studies of the Earth's field and its variations since little was understood of the field and its distribution across the world. For example, Sabine was instrumental in the studies of magnetism during British expeditions with James Clark Ross to the Canadian Arctic, where Ross located the North Magnetic Pole in 1831. Between 1839 and 1843, Ross explored in Antarctica where studies of the Earth's magnetism were part of the exploration objectives that also included geography, geology, botany and other subjects. Ross did not reach the South Magnetic Pole.

In the decades following Ross and his contemporaries Charles Wilkes of the United States and Jules-Sébastien-César Dumont D'Urville of France, expeditions to the Antarctic by many nations included scientific objectives related to studies of aurora and Earth's magnetism. Following the demonstration of long-distance communications across the Atlantic in 1901 by Marconi, it was clearly advantageous to investigate the propagation of radio waves in the southern polar region for science as well as for very practical – communications – objectives. The first attempt to use radio waves for communications in the Antarctic was in the Australian Antarctic Expedition of 1911–14 led by Douglas Mawson.

The International Geophysical Year (IGY) in 1957–58 ushered in intensive studies of the upper atmosphere and solar–terrestrial phenomena and processes, studies that have continued through to today with the employment of ever increasingly sophisticated instrumentation. Earth's ionosphere had been identified in the late 1920s as the reflecting layer in the high upper atmosphere that enabled long-distance wireless communications. Yet, as related by Dr Robert F. Benson (a winter-over scientist at the South Pole Station during the IGY year), at the time of the IGY it was unknown if the ionosphere would exist over the Pole during the long southern winter darkness and if communications could be maintained during this period. The ionosonde instrument used at the South Pole during the IGY demonstrated the existence of an ionosphere (albeit different than under sun-lit conditions), and showed that communications were therefore feasible.

This uncertainty at the South Pole in the IGY about the possibility of winter communications, because of a lack of knowledge of processes in Earth's space environment, is only one of many examples over the course of exploration of the Earth that illustrate the importance of the solar–terrestrial environment for enabling, facilitating, and even inhibiting the use of technologies. The effects of 'space weather' on technologies were first surprisingly encountered with the deployment of the electrical telegraph in the first half of the eighteenth century. At that time, spontaneous electrical currents were measured in telegraph lines in Britain, Europe and the eastern United States, currents that could readily disrupt telegraphed communications. As new electrical technologies continued to be developed, from wireless communications to power distributions systems to radar and navigation systems, the solar–terrestrial environment has always had to be factored into their design and operations. Thus studies of the space environment around Earth, and how the environment is affected by the Sun, are important not only for fundamental understanding but also for very practical reasons. The launching in the IGY of the first Earth-orbiting spacecraft, Sputnik, and the discovery of the Van Allen radiation belts with the launch of the Explorer 1 satellite, completely altered humankind's view of the Earth in space.

Figure 7.2 The aurora australis over the new South Pole Station. Auroras are produced by electrons (and protons) that strike the upper atmosphere. When oxygen or nitrogen atoms are hit by these energetic particles, they give off light at specific wavelengths (i.e. colours) as they return to their ground state. The green colour depicted above results from excited oxygen atoms between 110 and 250 km up. Although beautiful to observe, electrical currents that are produced during such auroral displays can have deleterious effects on a host of technological systems deployed on Earth and in space, such as satellites and power grids. Auroras occur almost each night in the auroral ovals that circle the magnetic poles and become visible at lower latitudes when the Sun is active and the solar wind is exceptionally variable. (Credit: Jonathan Berry, NSF)

Earth's space environment

Prior to the early space missions, the Earth's magnetic field had been viewed as essentially a large bar magnet – with North and South Magnetic Poles located in the southern and northern polar regions, respectively – sitting in the vacuum of space. The existence of the aurora and of cosmic rays (discovered in the first decade of the twentieth century) suggested to many early researchers that there was more to this simplistic bar magnet picture. The satellite era finally revealed that the magnetic field from the bar magnet was actually shaped roughly into the form of

Figure 7.3 An artist's conception of the interaction between the solar wind and Earth's magnetic field. (Credit: NASA)

an invisible comet – now called the magnetosphere – with the tail of the comet (the magnetotail) extending away from the Sun to distances perhaps as large as 1000 or more Earth radii. The front of the magnetosphere, facing the Sun, is defined by a shock wave that is formed when the solar wind (an ionised gas that is formed by the outward expansion of the Sun's hot (million and more degree) upper atmosphere, the corona, encounters the Earth's field. In physical terms, the pressure in the solar wind balances the pressure exerted by the magnetic field. This front boundary is at a nominal distance of about 10 Earth radii, and can move in and out under changing solar wind conditions driven by solar events.

Around the boundary and inside the magnetosphere is a complex environment of ionised gases, largely hydrogen ions and electrons, and electromagnetic and hydromagnetic waves of various types. The ionised gases are called 'plasmas'. There are also small amounts of ions of oxygen and nitrogen from Earth's atmosphere, as well as limited densities of ions from the Sun. The physical processes that control and change these ions and their distributions in space involve large-scale electric fields and Earth's bar magnet field. These electric fields, and therefore the ionised particle populations, can be drastically altered when solar storms occur and large amounts of the corona are expelled into the interplanetary medium – events called coronal mass ejections – and then encounter Earth. At such times, and even under less severe solar disturbances, Earth's magnetic field can be disturbed, with the resulting 'geomagnetic storms' – disturbances in the 'weather' of space – causing auroral displays and changes in the ionised particle distributions.

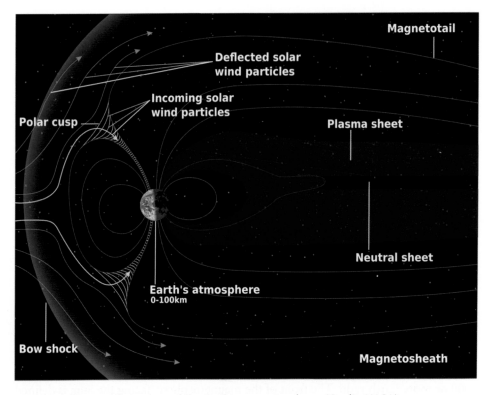

Figure 7.4 The overall structure of the Earth's magnetosphere. (Credit: NASA)

It is during such geomagnetic storms that disturbances and even interruption of technologies are observed.

Inside the magnetosphere are several distinct regions of ionised gas, even though these regions can be time-dependent in their structure depending upon solar conditions. At the very top of Earth's atmosphere, between altitudes of approximately 95 km and 250 km, is the ionosphere. This ionised layer around Earth is what enables wireless communications at low and high frequencies. Aurora occur in the ionosphere when the upper atmosphere is bombarded by charged particles from the magnetosphere and from the Sun under solar disturbances. Since the poles of Earth's dipole magnet are in the polar regions, charged particles from the sun, in the form of the solar wind and higher energies, can be funnelled into the polar caps along magnetic field lines that stretch upward to the boundary of the magnetosphere.

Two other major regions of ionised particles are the (1) plasmasphere, and (2) the radiation belts. The plasmasphere is a region extending to about three earth radii above the surface in which the ionised particles are largely from the ionosphere. The plasmasphere can expand and contract in size depending upon geomagnetic conditions as determined by solar emissions. The density of ionised particles

Figure 7.5 Above approximately 60 km, ultraviolet radiation and solar X-rays from the Sun ionise the upper atmosphere creating an ionised (charged) gas that is called a plasma. In the ionosphere, since individual particles are electrically charged, currents can flow and radio waves can be reflected enabling long-distance radio communication.

drops by a factor of about a thousand at the boundary of the plasmasphere – the plasmapause.

The radiation belts comprise regions of intense fluxes of much higher energy charged particles than those that exist in the plasmasphere. These high energy charged particles are confined by Earth's magnetic field. The intensities of these trapped particles can change with time under disturbed solar conditions, and it has even been found in the last decade or so that a third electron belt can occasionally form under very severely disturbed conditions. The process by which this can occur is unknown. Sensitive electronics in orbiting spacecraft can be affected, and even damaged, by radiation belt particles. The ionised particles in Earth's magnetotail can be accelerated in energy under disturbed solar conditions, and these particles can be driven into the upper atmosphere, causing and enhancing auroral displays. In addition, these accelerated particles can be added to the radiation belt populations.

Solar–terrestrial research in Antarctica

The Antarctic has a number of important advantages for studies of space-related phenomena, both solar–terrestrial research as well as astronomical research. In terms of solar–terrestrial research, several geographic and geomagnetic features make the Antarctic the first choice for important research. These include (1) the large land mass; (2) a wide expanse in coverage of both geographic and geomagnetic latitudes at all longitudes; (3) the foregoing item (2) derives directly from the fact that the highest latitude magnetic fields lines, those that can connect to the boundary of the magnetosphere and to the interplanetary medium, are measurable in the Antarctic while in the northern hemisphere this location is over the ocean; (4) the southern geomagnetic dipole on the land mass and the southern geomagnetic pole is very close to the land mass; (5) the high elevation of the South Pole (and other Antarctic locations) for optical studies (also applicable to some astronomical studies); (6) the wide range of areas

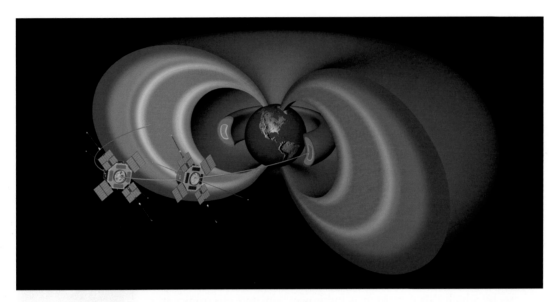

Figure 7.6 An artist's depiction of Earth's radiation belts. Particles in this region are very energetic and pose a danger to astronauts and spacecraft. Observations from numerous Antarctic locations help in understanding the dynamics of this region. The two satellites pictured are the NASA Van Allen Probes mission (launched August 2012) that, together with ground observations, are providing insight into the physical dynamics of the radiation belts. (Credit: Johns Hopkins University Applied Physics Laboratory)

in the northern hemisphere into which magnetic fields in the Antarctic can be 'mapped' (called 'conjugacy'; limited largely by the ocean coverage in the north); (7) large expanses of radio quiet areas.

Solar–terrestrial research in the Antarctic has evolved substantially over the 50 years since the IGY, moving from studies conducted largely at, and reported from, individual stations to the incorporation of measurements from multiple sites. Even at the time of the IGY and thereafter, studies at a single site generally involved instruments that measured different aspects of the solar–terrestrial environment. For example, at the South Pole in the IGY measurements included those of the aurora in optical wavelengths, of Earth's three magnetic field components, of the ionosphere and its layers, and of radio waves of various frequencies. Similar measurements are made today at many Antarctic locations, using much more sophisticated instrumentation. However, of more importance, these multiple location measurements are often combined together at the data acquisition stage and/or the analysis stage to provide essential information on spatial and temporal variations in the space-related phenomena.

The offset of the south geomagnetic pole (just offshore from the Dumont D'Urville station) and the south geographic pole (at the Amundsen-Scott South Pole

Figure 7.7 The wind and solar powered US Automatic Geophysical Observatory (AGO). These remote platforms on the Antarctic plateau consist of a suite of instruments used to monitor the space environment of Earth over a range of radio and optical wavelengths. Real-time data acquisition via satellites enables forecasting and now-casting capabilities for space weather applications. Although the AGOs were originally intended as space physics platforms (optical and radio wave auroral imagers, magnetometers and narrow- and wide-band radio receivers), the reliability and flexibility of the design has enabled other disciplines to leverage the location and infrastructure for new scientific instruments including seismometers, comprehensive weather stations and astrophysical sensors.

station) means that during every day a given location on the Antarctic continent will rotate through a range of geomagnetic latitudes. For example, South Pole station, at 90 degrees south geographic, has a geomagnetic latitude of about 74°S. Near local midnight, the South Pole will often be located on magnetic field lines that correspond to the auroral zone. Some 12 hours later, the South Pole will have rotated to be at local noon, and the magnetic field lines will now map to near the front boundary of the magnetosphere (where the field lines separate between those mapping to the front side and those mapping to the magnetotail). Other locations in Antarctica will similarly map out a range of geomagnetic latitudes during a local day. Thus, the rotation of Earth combined with the distribution of stations across

Antarctica provides important additional spatial information on space phenomena as mapped to Earth's surface. Of course, a similar situation exists in the northern hemisphere, but there is no land mass coverage across the North Pole for installing arrays of instruments.

A research problem of high significance, both scientifically as well as for more practical applications, is the state of the electrical currents in the ionosphere. These can be monitored across the Antarctic continent, and thus across many geomagnetic latitudes, through the employment of magnetometers at the manned stations of many nations as well as at many unmanned locations, such as those of the Automatic Geophysical Observatories of the United States and similar sites maintained by the United Kingdom and Japan. Combining these many magnetometer data sets and inverting them to produce ionosphere equivalent electrical currents yields maps of the spatial distributions of the currents and their changes with time.

These inferred ionosphere currents over the Antarctic continent can be 'mapped' along magnetic field lines for comparison with similar currents inferred from northern hemisphere measurements. Such mapping provides unique information on the global nature of Earth's ionosphere in polar regions as well as on the properties of the space environment at large distances from Earth (since the magnetic field lines 'map' to near the edges of the magnetosphere).

Another technique for measuring different aspects of the high latitude ionosphere uses spaced arrays of the order of six to eight radars (called SuperDARN radars) in both hemispheres. These radars measure the conditions of ionosphere plasmas and the measurements in the Antarctic can be again mapped into the northern hemisphere for inter-hemisphere comparisons and for high latitude global views of Earth's space environment.

Other instrumentation is also used at many of the magnetometer sites for measurements of other solar–terrestrial phenomena. For example, optical imaging systems at many sites are used during austral night-time conditions to map the changes in the aurora. These measurements are important in that it is not possible to get such widely spaced measurements of the aurora from the northern hemisphere, and certainly not at such high geomagnetic latitudes. And in conjunction with magnetometer and radar data, much more information on the state of the space environment is thereby obtained.

With so few people in the Antarctic, the radio quiet environment means that radio waves over a wide frequency range can be measured and monitored as a function of geomagnetic conditions and also in association with other measurements of the space phenomena. In particular, measures of radio waves at Very Low Frequencies (VLF; 3–30 kHz) are associated with important processes in the magnetosphere that can produce the loss of radiation belt particles. These

Figure 7.8 Ward Helms, left, and Michael Trimpi, right, record upper-atmosphere radio noise in a building constructed under the ice at Byrd Station during the International Geophysical Year. The Very Low Frequency (VLF) hut at Byrd station was a buried vault 2135 m north of the main station complex. These Stanford University projects were related to what would become the ongoing VLF projects that continue to this day at the South Pole and other sites across the Antarctic continent. VLF measurements are essential in remote sensing the ionosphere and magnetosphere. (Credit: NSF)

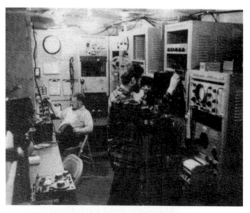

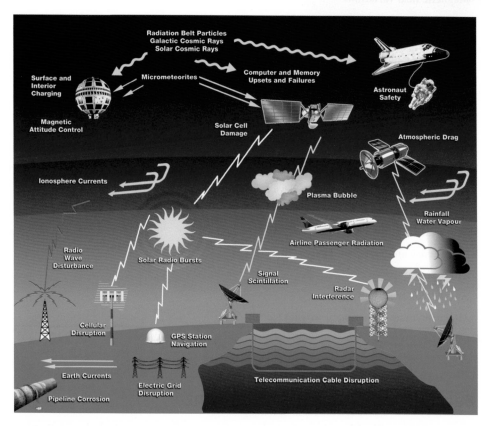

Figure 7.9 Space weather effects on technologies. (Credit: Alcatel-Lucent/New Versey Institute of Technology)

frequencies can be measured in the Antarctic at much lower signal intensities than generally possible in the northern hemisphere. In fact, the measurement of VLF waves in the Antarctic enabled the discovery of the plasmapause before spacecraft measurements were made of this magnetosphere feature.

Table 7.1 **Example space weather effects on technologies**

Ionosphere variations

Induction of electrical currents in Earth

Power distribution systems

Long communications cables

Pipelines

Interference with geophysical prospecting

Wireless signal reflection, propagation, attenuation

Communication satellite signal interference, scintillation

Magnetic field variations

Attitude control of spacecraft

Compasses

Solar radio bursts

Interference with radar

Excess noise in wireless communications systems

Charged particle radiation

Spacecraft solar cell damage

Semiconductor device damage and failure

Faulty operation of semiconductor devices

Spacecraft charging: surface and interior

Astronaut safety

Airline crew and passenger safety

Atmosphere

Low altitude spacecraft drag

Attenuation and scattering of wireless signals

Beyond Earth's space environment

Antarctica is an ideal location to conduct a myriad of astronomical observations, complementing the space and geophysical observations discussed above. Extremely cold, dry air and stable atmospheric conditions, together with the ability to observe objects continuously throughout the long winter or summer, offer unique conditions for observing our Sun, distant stars, supernova explosions and the cosmic microwave background. Today, astronomy in Antarctica is thriving, with major

Figure 7.10 The Cosmic Ray Observatory installation at Amundsen-Scott South Pole Station. Cosmic rays are subatomic particles that hit Earth's atmosphere and produce other subatomic particles, including protons, electrons, and neutrons, the latter of which can be detected at Earth's surface. The South Pole neutron monitor is perhaps the world's most sensitive detector of relativistic solar cosmic rays due to its location at high altitude and latitude. It is essentially closer to the space environment than most other monitors on Earth. By studying solar cosmic ray-produced neutrons scientists can monitor solar activity and cycles. (Credit: Paul Evenson, University of Delaware)

international facilities operated at many locations on the high Antarctic plateau. It is also interesting to note that Antarctica is a rich source for meteorites, providing astrobiologists with clues to the origin of life in the universe.

Some of the earliest accounts of astronomical observations south of the Antarctic Circle indeed date back to Cook's expedition, but those measurements focused on navigation accuracy. By most accounts, modern astrophysics was ushered in by the Australians in 1955, when cosmic ray detectors were installed at the recently completed Mawson Station.

The term 'cosmic ray' is really a misnomer, however, as these rays are predominantly energetic protons and helium nuclei. Showering down upon our planet, these particles are guided by Earth's magnetic field in much the same way that electrons of solar wind origin are guided to the polar regions, producing auroral displays. Entering the atmosphere above Antarctica, the protons strike air molecules, mainly nitrogen and oxygen, and produce a cascade of new secondary particles that include neutrons.

One of the early pioneers of cosmic ray research, and Antarctic astronomy in general, was Dr Martin Pomerantz of the Bartol Research Foundation. He conducted balloon-borne cosmic ray studies in the 1940s and 1950s and helped establish neutron monitoring sites at McMurdo Station in 1959 and at the South Pole in 1964. Neutron and cosmic ray monitoring has continued since those early experiments, providing one of the longest continuous data sets. The South Pole Air Shower Experiment (SPASE) in the mid-1980s continued the search for particles originating from violent supernova explosions. More recently, neutron monitors at McMurdo and Mawson stations operate as part of 'Spaceship Earth', a worldwide network of neutron detectors that provide real-time monitoring of cosmic ray distributions. Such measurements are critical for understanding high-energy particles coming

Box 7.1 **Helioseismology**

Helioseismology is the study of wave motions inside the Sun as measured from images of the solar surface. By monitoring these waves it is possible to obtain information about the solar interior, just as the study of seismic waves on Earth can provide insights into the structure of the interior of Earth. While the source of seismic waves is earthquakes, waves in the Sun are produced by motions in the convective region beneath the solar surface, an essentially continuous source. That is, when an earthquake occurs, the Earth can ring like a bell, struck once. In contrast, the Sun is like a bell that is being continually struck. Thus the solar waves are continuously present in images of the surface.

There are a number of different dominant wave modes that the Sun can oscillate in, and thus the measurement and analysis of images of the Sun can be quite complex. The desire is to have continuous 24-hour coverage of the Sun so as to be able to detect and measure solar modes over a wide range of their expected frequencies. Beginning in 1979, before there were continuous satellite images of the Sun over a 24-hour basis, the pioneering measurements by Martin Pomerance, Eric Fossat and Gerard Grec opened the South Pole to helioseismic studies. Observations from the South Pole in summer were used subsequently to good advantage to advance significantly the study of helioseismology. In particular, the South Pole measurements produced fundamental advances in understanding the internal sound speed and the internal rotational properties of the Sun.

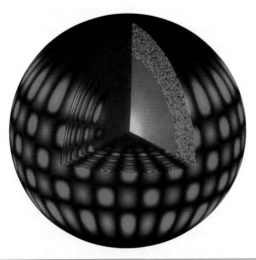

Figure 7.11 Waves resonating in the interior of the Sun. (Credit: National Solar Observatory)

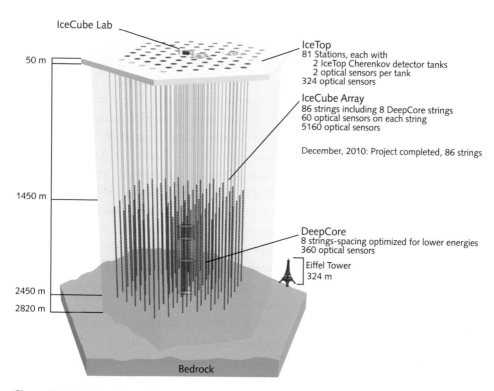

IceCube Lab

50 m

IceTop
81 Stations, each with
    2 IceTop Cherenkov detector tanks
    2 optical sensors per tank
324 optical sensors

IceCube Array
86 strings including 8 DeepCore strings
60 optical sensors on each string
5160 optical sensors

December, 2010: Project completed, 86 strings

1450 m

DeepCore
8 strings-spacing optimized for lower energies
360 optical sensors

Eiffel Tower
324 m

2450 m
2820 m

Bedrock

Figure 7.12 Artistic view of the IceCube neutrino telescope with the array of photomultiplier tubes buried in the ice under South Pole Station. (Credit: Danielle Vevea/NSF and Jamie Yang/NSF)

from the Sun and for space weather forecasting, such as determining radio communication interference.

High-energy detectors are also used to search for elusive particles called neutrinos, believed to emanate from some of the most exotic astrophysical objects such as quasars and black holes. In the early 1990s, the AMANDA (Antarctic Muon And Neutrino Detector Array) experiment was deployed at South Pole to search for these particles. The experiment consisted of widely spaced photo-multiplier tubes (PMTs) buried in the ice and facing downward, thereby using Earth to filter out most other particles, including cosmic rays. Essentially, the detector was the large chunk of ice outlined by the PMTs. Since neutrinos barely interact with matter, they can usually travel directly through Earth unscathed. However, when the occasional neutrino interacts with the ice, a blue flash of light known as Cherenkov radiation is produced, which can then be recorded by the PMTs. The next generation neutrino telescope was recently completed at South Pole.

Coined IceCube, it is literally a kilometre by kilometre detector buried in the ice under South Pole.

In the 1990s, astronomers further recognised that the cold, dry atmosphere of the Antarctic plateau provided several excellent viewing windows in the electromagnetic spectrum in which to observe the Universe. These included regions in the microwave, sub-millimetre and infrared parts of the electromagnetic spectrum. A concerted effort to organise the astronomy community was initiated at South Pole under the National Science Foundation's Center for Antarctic Research in Astronomy (CARA), and one of the first instruments installed was the South Pole Infrared Explorer (SPIREX). This 0.6-m diameter SPIREX telescope was designed to observe cool stars, distant galaxies and heavily obscured star-forming regions. Serendipitously, however, during the long winter night in July 1994, the winter-over research crew captured some of the world's best pictures of the collision between the comet Shoemaker–Levy and Jupiter. South Pole was uniquely positioned to observe this collision since Jupiter did not rise or set at this location, enabling continuous observations.

Astrophysicists in Antarctica have also turned their attention to fundamental properties of the cosmic microwave background (CMB), the afterglow of the Big Bang that was discovered in 1965 at Bell Laboratories by Arno Penzias and Robert Wilson. Using a radio telescope called the Degree Angular Scale Interferometer (DASI) at South Pole Station, University of Chicago scientists measured a slight polarisation to the CMB. Earlier results revealed small temperature differences in the cosmic microwave background, supporting a dramatic expansion explanation of the early Universe known as inflation theory. Several new initiatives include the South Pole Telescope, a 10 m sub-millimetre telescope designed to investigate the properties of the mysterious dark energy that permeates the Universe.

At even higher altitudes on the plateau, there is promise for perhaps even better seeing conditions than at South Pole. A French–Italian station at Dome C is now complete and operating through the winter and site testing is underway through the Concordia-astronomy programme. At Dome A, the top of the Antarctic plateau, China has established a new station, and deployed the Australian-built PLATeau Observatory, a truly international collaboration, with instruments contributed from Australia, China, New Zealand, the United Kingdom and the United States. One of the initial instruments deployed is CSTAR (Chinese Small Telescope ARray). This telescope is composed of four identical telescopes that will study variable stars and sky brightness. The newly established Astronomy and Astrophysics research programme under the auspices of the Scientific Committee on Antarctic Research should enable and foster such collaborative efforts for the foreseeable future.

To the future in space science research

Research in the space sciences in the Antarctic has evolved over the decades following the IGY from vacuum-tube instrumentation at single sites devoted to purely Earth-centred projects, to highly coordinated arrays of instruments spread over the continent and to the most sophisticated astronomical telescopes. Data from these instruments are now widely used to develop new models and new theoretical understandings of Earth's solar–terrestrial environment, of the local Milky Way galaxy, and of the universe.

It is difficult, and indeed foolhardy, to predict where research directions and emphases will trend in the next decade – and certainly in the next five decades leading to the next 'polar year' (or if there will even be some other emphasis on polar research at that time). Certainly, given the planning and logistics required to successfully carry out research in the hostile climatic environment of the Antarctic continent, research over the next half decade or so will continue to exploit the instrumentations (including astronomical telescopes) that are currently in place and are planned. The arrays of instruments from many nations will increasingly be linked together to obtain a more detailed and real-time perspective of the solar–terrestrial environment. There will also be new instruments developed for upper-atmosphere studies over the next decade. These instruments will be more sophisticated, especially optical instruments with new imaging and spectral resolution characteristics.

The astronomical telescopes and their data systems at South Pole, Dome C, and other possible future locations require high band-width communications to return data to home institutions. This enhancement in communications is a major technological (and hence cost) challenge for Antarctic-support nations. Bringing the full astronomical databases back to home institutions will allow results to be achieved more quickly and will permit more real-time adjustments in observing schedules. More remote handling of the telescope observing schedules and operations will also be possible. Higher band-width communications will also enable the solar–terrestrial research programmes to develop innovate and adaptive research operations that can respond to real-time events and contribute more broadly to space weather forecasting.

The next decade will also see more technological developments in the use of wind and solar power at Antarctic locations at year-round, seasonal field, and remote unmanned locations. Together with a continued decrease in electrical demands from individual instruments due to innovations in electronics parts and designs, clever new station power source designs will enable a much more vigorous and expansive deployment of remote observation sites. Such a development will make even better use of the Antarctic landmass for remote station deployments in order to gain deeper understanding of solar–terrestrial relations.

In summary, the future of space research in the Antarctic – solar–terrestrial and astronomical – is quite bright, with many ideas and innovations in progress, in planning and on the horizon.

8 Living and working in the cold

LOU SANSON

Adventure is just bad planning. **Roald Amundsen**

Not many people get to work in Antarctica and, for many of those that do, it can prove a life-changing experience. Every year over 2000 scientists and support personnel travel south to conduct scientific research. They come from over 30 countries and, while some stay only a few days, others stay for months or even a year or two. Some spend their time at the 40 permanent research stations scattered around the continent that are open year around; others make use of the 30 stations that are used in the summer only or undertake long research cruises offshore, and many more set up field camps deep in the most remote parts of the continent.

The modern research stations have little in common with the historic huts of Robert Falcon Scott and Ernest Shackleton. Built in the early 1900s, their wooden buildings provided little more than shelter from the cold and wind and a place to wait out the winter. But today's stations, from the smallest, like Norway's Tor (home to only four), to its largest, the United States' McMurdo (where over 1000 live in the summer), Antarctic research stations are like self-sufficient villages. Each needs facilities for power generation and for dealing with waste, for making fresh water and for medical care. The stations must be able to store and prepare food and provide accommodation for scientific and support personnel as well as be capable of maintaining and servicing vehicles and running telecommunications units, and often maintain sophisticated laboratories.

Geology, surveying, meteorology and some biology were being undertaken on the Antarctic Peninsula after the World War II by the UK, Chile and Argentina. Whilst the French built a station at Port Martin in 1950, they had hardly started their science before the station burnt down. So the first major research facility

Antarctica: Global Science from a Frozen Continent, ed. David W. H. Walton. Published by Cambridge University Press. © Cambridge University Press 2013.

Figure 8.1 Jamesway huts at McMurdo in the 1950s. (Credit: AntarcticaNewZealand)

in East Antarctica and a precursor of the modern science stations was Mawson, constructed by the Australians in 1954. Mawson station led the first surge of building that occurred largely in response to the International Geophysical Year (IGY), and its opening was followed in December 1955 by the United States' McMurdo Station on Ross Island and then the Amundsen–Scott South Pole Station in 1956.

By 1958, 12 countries had constructed stations around Antarctica. The Americans relied initially on Jamesway buildings, similar to military Quonset huts but with wooden frames covered by insulated fabric (typically canvas), and have continued to use them since for field camps. The huts can contain sections for sleeping, eating, washing and working and are often kept warm by oil-burning heaters, while propane stoves are used for cooking. Another popular building material during the early construction of stations was polyurethane panels, like the ones used in the manufacture of food freezers. However, unlike in food freezers, when used to build an Antarctic field station, the panels keep the cold out and the warmth in. Many other stations were constructed of wood with heavily insulated cavities between the walls. The Logistics Working Group of SCAR provided an important early forum for technical discussions on buildings,

Figure 8.2 Insulation panels in the buildings at Scott Base. (Credit: AntarcticaNewZealand)

both in terms of style and structure. One perennial problem was caused in many sites by drifting snow, which gradually overwhelmed the buildings. For this reason, and for ease of getting supplies and people in, windy rocky places on or near the coast were usually the first choice of site as they posed fewer problems than building on snow or ice.

Many of the present permanent research facilities have been constructed since the late 1980s. Many of the first stations have been buried by snow, as have their replacements. The British station at Halley is now in its sixth version, with the others all abandoned as they became either uninhabitable (squashed by the weight of snow) or unsafe as they moved out with the ice flow onto a floating ice shelf. The original structures at Amundsen–Scott South Pole Station were abandoned in 1975. Upgrades and new construction have been installed that make the latest facility there as close to being in a space station as anywhere on Earth.

Construction in Antarctica needs to be adapted to the harsh environment – this is definitely a challenging place to build. For example, structures that are built on top of ice or snow produce heat and can slowly sink if not well insulated. On the other hand, drifting and blowing snow can bury buildings on the surface and not only must pipes be protected from freezing but, with wind gusts over 150 knots not uncommon, doors and windows can blow in, filling buildings with snow.

There is currently enthusiasm for remodelling and rebuilding stations using modern engineering and design features such as raised platforms and legs that can be jacked up as snow accumulates. For example, Germany's Neumayer Station has now been replaced by a two-storey building that is built on hydraulic piles. The new station – called Neumayer III – was prefabricated in Germany, sent by ship to Antarctica, hauled by snow tractor to the research site at Atka Bay on the Weddell Sea and assembled on site to start functioning in 2009.

The British Antarctic Survey – responsible for sending 450 scientists and support personnel to the Ice annually – has also built an advanced new station, which it commissioned in 2012. The Halley VI modules are on platforms on top of a series of 'mechanical legs' that are mounted on skis. These legs allow the structure to be towed by tractors to a new site inland as glacial calving and snowfall demand. This futuristic station has little in common with the original buildings of the 1950s.

Figure 8.3 South Pole Station after rebuilding in 1975 with a geodesic dome. (Credit: AntarcticaNewZealand)

The north platform provides year-round accommodation, which is supplemented by summer living space in the south platform (capacity is 16 in the winter and 52 in the summer) that also houses the science facilities. The central module is designed for socialising and recreation and contains a dining room, gym, pool, sauna, hydrotherapy bath, a double-height climbing wall and facilities for music and arts and crafts. It also has panels that change colour in the winter to mimic the changing light in a temperate climate.

Construction of the buildings at the permanent stations was originally purely functional – they were designed to provide protection from the elements and a place to work, sleep and eat. There was no concern for the interior comforts of the buildings – after all they were meant to house explorers! Sadly, little concern was also paid to the long-term environmental implications of human presence on the continent. It is hard to imagine now, but 50 years ago, the snow and ice-free areas of Adélie penguin colonies seemed like perfect places to build research stations. Thankfully, a lot has changed. In 1991, the Protocol on Environmental Protection was added to the Antarctic Treaty and since then the focus of designing stations – and the activities at these – has shifted to one which has minimum impact on the environment.

The high costs of building and maintaining stations in Antarctica might have been expected to encourage the development of international shared facilities. Whilst many stations support visiting scientists from other countries only two could be described as international – Concordia run jointly by France and Italy, and Jubany, an Argentine station with a major German biological laboratory attached to it. One country, the Netherlands, has decided not to build a station but to rent space and facilities from countries that have spare capacity – a more efficient use of resources and another way of minimising impacts.

Figure 8.4 The new South Pole Station commissioned in 2008. (Credit: Robert Schwarz, NSF)

Figure 8.5 (right) Computer graphic of the structure of Neumayer III station showing the hydraulic columns that keep it above the snow surface and the undersnow garage. (Credit: Alfred Wegener Institute for Polar and Marine Research)

Figure 8.6 (above) The new Halley VI station on the Brunt Ice Shelf, made of modules mounted on skis so that the station can be towed to a new site. (Credit: BAS)

Figure 8.7 (right) Concordia station, run jointly by France and Italy. (Credit: Y Frenot, IPEV)

Box 8.1 **Dealing with wastes**

Most activities carried out in Antarctica will produce waste that can be harmful to the Antarctic environment. Certain products are prohibited from being brought to Antarctica, including PCBs, pesticides, polystyrene beads or chips and live animals. Research programmes today are much more considerate to the environment than when the first explorers and scientists travelled to Antarctica. Back then, for example, human waste was simply put as raw sewage into the sea – today, on many stations, it undergoes biological and UV treatment in sealed effluent plants. The discharged effluent meets World Health Organisation standards and solids are transferred back to the host countries for destruction. Other waste has to be sorted so that it can be recycled, then it is normally crushed, containerised and removed back to the host countries. Managing the waste streams is a continuing activity for the station environmental officer.

Figure 8.8 Waste at Halley is compacted and packed into containers for shipment back to the UK. (Credit: Chris Gilbert BAS)

Energy-efficient upgrades are being made to existing stations and many countries are taking advantage of developments in solar and wind power. Significant breakthroughs in the construction of new stations have been seen during International Polar Year (2007–09). One of the most noteworthy comes from Belgium. This country, which closed its first science station in 1967, now has a new station, Princess Elisabeth, in the Sør Rondane Mountains in Dronning Maud Land. This has proved innovative in many ways. It is the first station to have zero emissions and is powered completely by renewable energy sources – wind turbines and solar panels – managed through the novel concept of a smart grid. This is quite a feat, as fuel demand on stations is high – not only because they must be heated year round, but power is needed for making fresh water, managing the sewage and waste facilities as well as powering the science equipment. As it is only inhabited during the summer, it is managed remotely during the winter through a communications link. Designed and built by the International Polar Foundation (IPF) using mainly private funding it was donated to the Belgian Government in 2010 with IPF as the mandated Antarctic operator for the Belgian state.

Figure 8.9 Princess Elisabeth station showing the aerodynamic design of the station and the commitment to sustainable energy through wind turbines and solar panels. (Credit: Réne Robert, International Polar Foundation)

Construction of the station is estimated to have cost around 22 million euros, and will house 40 people over summer periods.

The designs of Princess Elisabeth, South Pole or Halley VI are about as far from the early huts of Scott and Shackleton as imaginable. Life inside them is dramatically different, too. At South Pole, for example, there is an area where vegetables can be grown hydroponically. The early explorers would probably have never imagined eating fresh vegetables whilst living in Antarctica! Realistically, even today for much of the year there are not very many available for the majority of researchers and support crew. During the summer, even regularly scheduled deliveries by ship or airplane can be delayed because of bad weather. In the winter, there are no shipments, so food is frozen, tinned or dried. Because the Antarctic Treaty forbids bringing soil to Antarctica, greenhouses are not an option for growing plants but hydroponic systems – where plants are grown in slowly circulating nutrient solutions – are used at a few of the larger, permanent stations like McMurdo, South Pole and the Australian stations.

Figure 8.10 Hydroponics in the South Pole Food Growth Chamber, which is a favourite place for people to go and relax. (Credit: Lane Patterson, NSF)

At the smaller stations, scientists and support personnel typically share the responsibilities for cooking, but most of the larger facilities have professional chefs. It takes a lot of energy to fuel bodies working outside in extreme cold and menus are typically developed by a nutritionist or food management specialist to be nutritionally competent and provide a variety of interest. Food is basically the same as you would eat at home and menus reflect the cuisine of the home country; for instance, you might find marmite on toast for breakfast at New Zealand's Scott Base and pasta and wine for dinner at Italy's Mario Zucchelli Station. The pantry at the German station is stocked with tins of sauerkraut, there is plenty of rice at India's Maitri Station and, what else but turkeys are roasted for Thanksgiving dinner at the American's McMurdo.

Most stations have a limited supply of alcohol and when the weather is nice, social activities might include barbecues or cross-country skiing; many of them also have indoor gym equipment and most have libraries and room to play music, pool or board games or watch DVDs. The largest station, McMurdo, even has a small bowling alley!

Packaging of foodstuffs (and other supplies) must be kept to an absolute minimum and nothing can be wasted. Meat is brought to Antarctica boneless and, just like in a practical home kitchen, leftovers are creatively used in the next day's meals. Whilst on many stations everyone helps out with the cleaning up, some stations now employ contractors to undertake all the service support.

For stations near the coast, fresh water is made by desalinating seawater through a reverse osmosis process. At inland stations, ice and snow are melted to make fresh water. Generators supply power and heat to most of the stations and the majority are still fuelled by diesel. Historically, this has been a logical power source, but it comes at a high financial and environmental cost – the implications of fuel spills and damage to the atmosphere are most extreme on the Antarctic continent. Just like everything else, fuel needs to be taken to Antarctica by ship or airplane and then transported to the more remote localities. The installation of wind turbines by New Zealand and Australia is generating wider interest in replacing at least some diesel power with more sustainable power sources.

Depending on their home country, scientists and staff arrive by sea or air and many national programmes work cooperatively to make transportation as economic

Figure 8.11 Field fuel dump, which often gets covered by blowing snow and has to be dug out by hand. (Credit: Chris Gilbert BAS)

and efficient as possible. For example, the New Zealand, United States and Italian programmes are all based in Christchurch, New Zealand, and share a joint logistics pool with each country contributing airplanes, equipment and personnel. In the USAF C17 the trip to Antarctica from Christchurch takes approximately 5 hours; in the RNZAF Boeing 757, it is around 4.5 hours whilst in a Hercules it is over 8 hours. The airplanes land on a sea ice runway at McMurdo from October to December. After the ice begins to soften and melt in mid-summer, the flights land at Pegasus white ice runway about 35 km from the stations. In addition to the air movements, a yearly ship-based resupply of fuel and cargo to McMurdo, Scott Base and Mario Zucchelli Station is supported by icebreakers.

Running a joint logistics pool has been replicated in South Africa where 11 countries have pooled their resources to create the Dronning Maud Land Air Network (DROMLAN) with commercial aircraft. Flights leave Cape Town International Airport and fly to the 3000-m ice runway at Novo Air Base close to the

Figure 8.12 RNZAF Boeing 757 on the Pegasus white ice runway at Williams Field. (Credit: AntarcticaNewZealand)

Russian Novolazarevskaya Station and serve scientists and personnel from Belgium, Finland, Germany, India, Japan, the Netherlands, Norway, Russia, South Africa, Sweden and the United Kingdom.

Ships are used by many programmes including the British Antarctic Survey, which runs two ice-strengthened research ships and also uses a navy ship for survey work and helicopter support. Most of the South American countries use navy ships for support. The most active Antarctic countries such as Australia, South Africa, France, Germany, Japan, Korea, Spain, the United States and China also have dedicated Antarctic research ships, some of them capable of ice breaking. Until 2007, Australia had a purely ship-based programme with 30-day long trips setting sail from Hobart. The Australian Antarctic Division has recently contracted with a commercial airline to transport scientists and support crew on an Airbus 319 that flies from Hobart. The three Scandinavian countries – Norway, Sweden and Finland – have always closely cooperated, sharing the logistics costs of ships, and more recently the tourist cruise ships have also assisted several countries by delivering staff and resources to several stations during their summer visits.

Land travel is often by snowmobile or special polar tractors like Hagglunds and Pisten Bullies. Marker poles indicate safe travelling routes, but when crossing infrequently travelled areas, parties must stop regularly and drill through the ice to measure its thickness. Deep and deadly crevasses can be covered by a thin layer of snow or ice which suddenly gives way. This sort of travel is only possible with highly skilled field guides.

If travelling around Antarctica required battling only extreme cold and wind, it would be challenging enough to keep vehicles running; but add to its extreme weather the mountainous terrain, crevasses and melting sea ice and you have an environment that makes transport nearly impossible. Keeping equipment running in such an environment is tough. Not only is metal prone to cracking and breaking

Box 8.2 **Antarctic aircraft**

Arriving in Antarctica by air rather than by ship is increasingly common as operators try to make the most use of the short summer season. All the intercontinental aircraft other than Hercules are wheels only and so have to land on gravel, sea ice, compacted snow or blue ice runways. At present there are only three gravel runways at Rothera, Marsh and Marrambio. Once in Antarctica, getting to a deep field camp may require air transport by helicopter or fixed wing aircraft and these can only operate during the summer. Helicopters are often used from supply ships but at McMurdo/Scott Base and at Mario Zucchelli commercial helicopter companies provide summer facilities. There is a very limited range of fixed wing aircraft certified to operate on both wheels and skis so many programmes use ski-fitted Twin Otters or the Basler DC-3. Typically, aircraft are ferried down at the beginning of the season along with pilots, engineers and mechanics although aircraft have overwintered on the continent either in hangars or carefully buried under protective snow. Pilots undergo specialist training, and the aircraft carry first-aid kits and emergency camping equipment. Bad weather can often ground flights headed to Antarctica as well as flights around the continent. There are several wheel only aircraft that operate into the interior and these rely on landing in blue ice areas where the snow has been completely blown away and a surface of polished ice is the runway. These need to be long and flat as the pilots cannot use brakes to slow down on the ice and must use reverse thrust instead.

Figure 8.13 (opposite) Hard rock runway at Rothera, which provides a base for Twin Otters and a DASH-7 during the summer. With hangar facilities and a fuel farm this British airstrip is used by a number of national operators every season. (Credit: Peter Bucktrout, BAS)

Figure 8.14 (top right) Twin Otter showing ski/wheel configuration. (Credit: AntarcticaNewZealand)

Figure 8.15 (centre right) A DC-3T Basler operated by Ken Borek Air Ltd out of Calgary in Canada for the US Antarctic Program. (Credit: Kevin Bliss, NSF)

Figure 8.16 (below) Three Twin Otters on the blue ice runway at the SkyBlu refuelling station. (Credit: D. W. H. Walton, BAS)

Figure 8.17 Haglund tractor train.
(Credit: AntarcticaNewZealand)

in temperatures that drop to –60°C (−76°F) but fingers do not work too well in
the cold either, with the skin sticking to cold metal if you are not very careful. Most
of the research stations have garages to bring equipment in for repair, but if in the
field, what might take a couple minutes inside can turn into a day-long project.
Hands get cold and lose their dexterity and it is impossible to be precise wearing
big mittens and gloves.

Preparing for fieldwork is a lot like planning a camping trip to outer space or
going on an extended sailing voyage – absolutely everything must be taken along and
provisions for emergencies must be included. When storms come up, there is no way
even the most severely injured or ill personnel can be rescued, so each field party
must be outfitted with a first aid kit, emergency supplies and the knowledge and
confidence to use them.

Access to deep field camps is usually by helicopter or fixed-wing aircraft and
both have space and weight restrictions, making packing and planning a very
important part of the process. In general, it takes about a year to prepare fully
for a deep field trip. Scientists are typically required to procure all the scientific
staff and equipment they will need to conduct their work. Equipment must be
able to withstand the travel and the extreme conditions. Batteries to power
photographic equipment and communications equipment need to be kept
warm in order to work – consequently equipment is often tucked snugly inside
clothing during the day and sleeping bags at night.

Camping in the field is not glamorous and toilet facilities usually consist of a
20-litre plastic drum inside a tent or behind a snow wall. Waste is collected and
taken back to the main station where it is processed through the sewage treatment
plant. Women often use a 'Female Urinary Device' or FUD, which is a soft plastic
funnel apparatus that slides inside clothing and fits snugly against the skin.

Box 8.3 **Huskies**

The first explorers brought horses and dogs. The terrain proved difficult for horses but dogs were used extensively by Amundsen to get to the South Pole in 1911 and in the 1950s and 1960s for virtually all of the Transantarctic Mountains expeditions and the exploration of the Antarctic Peninsula. By the 1970s most countries had begun to replace dogs with snowmobiles. While dogs are an important part of history, keeping non-native animals on Antarctica was banned because there was the possibility that canine distemper disease, a serious problem in the Arctic, could spread to Antarctica's seals. The other concern was that, if dogs broke loose, they could attack wildlife. The strict environmental regulations that most national programmes now follow include removal of all waste material and it would be nearly impossible to maintain dogs to that standard. The British, Australian and Argentine programmes continued to keep dog teams until they were all removed from the continent in 1994 – signaling the end of the historic era.

Figure 8.18 Husky team under Mt Ole Engelstad, NZARP Southern Party 1961–62. (Credit: Peter Otway, Antarctica NZ)

Box 8.4 **Field accommodation**

Most Antarctic science takes place in field locations far from the permanent stations and the comforts they contain. The type and standard of the accommodation varies widely. Around much of the Peninsula and in some East Antarctic locations there are fixed field huts made of containers, wood or converted vans. These are usually installed to allow regular work at particular sites such as lakes or penguin rookeries. These often have food stocks, fuel and bedding and can also function as emergency shelters for parties that are caught out in bad weather. Many of them double up as both accommodation and field laboratories. For larger parties and longer term work, Jameswaysare often used by some operators as are other permanent, seasonal facilities in locations such as Bratina Island where staff stay for weeks or months. Most short-term field accommodation is based around the pyramid-designed canvas tents similar to those that Scott used in 1902 and this is especially suitable for small parties undertaking traverses or survey work. Travelling with skidoos and sledges these small parties are normally put into the field by air and then can remain self-sufficient often for many weeks. The more modern 'Rac-tent' is designed to make most use of the interior space and can be supplemented by lightweight fibreglass Apple huts that are positioned by helicopter.

Figure 8.19 (opposite) Scott tents in the Taylor Valley, Victoria Land. (Credit: Chris Kannen, NSF)

Figure 8.20 (this page, top) Rac-tents. (Credit: AntarcticaNewZealand)

Figure 8.21 (this page, bottom) Apple hut for longer field party stays. (Credit: AntarcticaNewZealand)

Figure 8.22 Heroic Age clothing from Shackleton's Ross Sea Shore Party in 1916, still hanging in the hut at Cape Evans. (Credit: Ralph Maestas, NSF)

People working in the field tend to do more physical work than those at the stations and burn lots of calories just keeping warm. Consequently, food needs to be high in energy, easily cooked and requiring little fuel. Dehydrated food makes up a big proportion of meals eaten in the field and ration boxes contain a variety of groceries including chocolate, rice, biscuits, muesli bars and dried milk, fruit, soup, meat and vegetables. As backcountry food in general has improved, there has been an increasing shift to providing more high-nutrition rations that are tasteful and interesting. Water is obtained by melting snow over a gas- or paraffin-fired primus stove and, in order to save fuel, meals need to be quick and easy to prepare. Even though most of the larger camps will set up a tent to be used as a kitchen and dining hall, bad weather can force people to stay inside their tents for days or even weeks on end. Consequently, everyone needs to have emergency food, water and stoves inside their tent. In huts and tents, a synthetic camping mat or air mattress is used and two or three huge down-filled sleeping bags (with hoods) are stuffed together and campers typically sleep in their clothes when it is really cold. Washing, both of the body and of clothes, is clearly difficult so long-stay field workers tend to smell rather strongly when they get back to the station!

Staying warm and dry is the first step in being safe and polar gear is specifically designed for the hostile conditions of Antarctica. A lot more is known today about

Figure 8.23 Erik Barnes (left) wearing New Zealand Antarctic clothing and Brian Stone (right) with US clothing. (Credit: Chris Demarest, NSF)

keeping the body warm than 100 years ago when the first explorers dressed in cotton and wool. An often-used guideline is that you should never dress or work so that you perspire – becoming wet with sweat is a sure way to become chilled, especially if the sweat freezes! When perspiration occurs, the clothing needs to draw that moisture away from the body. Most programmes provide several layers of lightweight clothing instead of one or two layers of bulky, thick gear. Layers provide air for insulation and can be added or removed depending on the temperature – they also allow the wearer to move freely.

A typical clothing 'kit' that scientists and support personnel take to Antarctica weighs about 20 kg and consists of a multi-layer system of polypropylene, down and thinsulate. Footwear includes double insulated mountaineering boots for moderate temperatures and conditions, and knee-high muk-luks for extremely cold weather. Layers of highly insulated socks are worn and for those who need them, warmer packs are provided which are placed between the skin and socks; when broken, the packs provide extra warmth. Fingers are prone to frostbite and two or three layers of gloves are worn. The heat packs can also be used inside gloves. The intense sunlight and glare of the snow is hard on the eyes and sunglasses or ski goggles are worn at all times outdoors.

As a safety precaution, the buddy system is used when venturing away from a station, as is a sign-out procedure, which includes details of destination, time and date of departure and expected time of return. Most bases require that every party leaving the station carries a hand-held VHF radio. Being alone in the Antarctic is difficult!

When proper precautions are taken, Antarctica is a very safe place to work with a very low mortality. Clothing, eye protection and sun block are used to protect workers from the strong UV rays especially in spring. Though protection is the best defence, frostbite and hypothermia are very real risks in Antarctica. Every day bumps and bruises, sprained ankles and cut fingers are just as likely to occur in Antarctica as anywhere else, but if a person becomes seriously ill or injured, it is often impossible to evacuate them. This is especially true when the weather goes sour and for deep field parties and those staying over the winter. Medical care in Antarctica became front page news in 1999 when Jerri Nielsen, the only doctor wintering over at the United States' South Pole Station, diagnosed and treated herself

Box 8.5 **Diving**

For those stations that run marine biological programmes there are special considerations. At McMurdo, Rothera, Casey and Mario Zucchelli there are regular diving projects that mean the stations need decompression chambers and their doctors need expertise in diving medicine. Each station needs an experienced diving officer to manage the activities whilst the scientists and technicians need to learn new rules. Divers may enter the water by walking in off a beach but more commonly they travel to dive sites in a small boat or on a sledge towed over sea ice. They usually wear a dry suit with several layers of clothes beneath it, a hood and several layers of gloves. Modified regulators are used to avoid the gas lines freezing up. Holes are cut in the ice with chain saws and in some places kept open for long periods with a hut over them. Often the coldest part of the dive is getting to and from the site – in the water the temperature is only around $-1°C$)!

Figure 8.24 Decompression chamber at Rothera. (Credit: Chris Gilbert BAS)

Figure 8.25 Science research divers Dug Coons and Henry Kaiser at New Harbor, diving through ice to collect specimens. The hole in the sea ice is normally cut with a chainsaw and needs to be continually 'fished' to keep the ice crystals from coalescing and sealing the hole. (Credit: Henry Kaiser, NSF)

for an aggressive form of breast cancer. Her story brought to sudden focus the challenges faced by everyone who travels to and works on the Ice.

Prior to travelling to the Ice, all personnel must pass rigorous health screening that includes chest X-rays and tests for diseases and conditions such as hepatitis and HIV. A doctor is in residence at most stations, certainly during the summer, and most have facilities to conduct a wide range of basic treatments. Some, like the Chilean Teniente Marsh and the Argentine Marrambio stations, have small hospital facilities, not least because they also have to provide medical support for the families who live on these stations.

Just like the rest of the world, communication in Antarctica has dramatically changed in the last 50 years and most stations have internet and telephone services allowing doctors to use telemedicine techniques for both diagnosis and treatments. Remote health care has now reached a very sophisticated level with station doctors able to discuss cases directly with consultants elsewhere in the world and send back digital X-rays or ECGs for diagnosis purposes.

Scientists and staff who go to remote field stations communicate back to their Antarctic home base either by high-frequency radios or increasingly by email. Field parties normally have a scheduled time to make radio contact and report on the weather, progress and any problems that might have arisen. Satellite phones can be taken to the field and pre-paid phone cards as well as email are the way most scientists and staff communicate with their families back home. You can now surf the internet from many Antarctic stations and the real problem has become communications bandwidth on the satellites as people begin to expect the same level of connectivity that they have at home.

Choosing staff to winter-over – or even work through the summer – is not a science, but it definitely requires both experience and attention to detail. A big part of the decision on who to hire is based on how individuals will fit together in the overall team. The system for selecting staff varies widely across countries – some use psychological testing, others do not; some draft in military personnel, others employ only civilians.

Of course, candidates who want to work in the Antarctic need to be skilled in their specific area; whether they are carpenters or chefs, microbiologists or engineers, those chosen need to be experienced and flexible. But on top of the 'work experience', people need to be emotionally intelligent, tolerant and self-aware, able to deal with conflict, stress and feelings of homesickness.

People who are most successful working in Antarctica are enthusiastic about their work, but are laid back in their approach to life. Obviously, everyone has different values and morals and in a normal working situation, people do not see their co-worker's personal life in such detail. But at a small research station, living cheek by jowl, there are no secrets. Antarctica is an isolated place, far from friends,

Figure 8.26 Digging a snow hole near Scott Base as part of the field training. (Credit: AntarcticaNewZealand)

family and support groups and a lot of the support personnel – for example, housekeepers and chefs – chose to live a transient lifestyle and are accustomed to moving around and meeting new people.

Often, people see working in Antarctica through rose-tinted glasses, but the reality is very different and it is usually hard work. All of the support jobs are about customer service and even the most self-sufficient person will miss home, family and friends. In the summer, life in Antarctica is really busy – there are a lot of people around and with 24 hours of daylight, activity can go on around the clock. But there are still times when you cannot go outside. During winter storms, that is even more likely and winter is a tougher time to live there. Not only is it dark, but there are fewer people around, much less fresh fruit and vegetables, and even if there is an emergency, evacuation is not normally possible. Consequently, selection of wintering staff needs to be even more careful.

Most programmes invest quite heavily in pre-deployment training where staff learn about the Antarctic Treaty and why they were employed and what to expect. Some roles, for example, science technicians, spend months learning the job, the

Box 8.6 **Women in Antarctica**

Whilst the early Antarctic expeditions consisted only of men, it began to be clear in some countries after the IGY that women could play a productive part in Antarctic work. Initially, it was the United States that began to take female scientists down in 1969 when a party of six worked at McMurdo and visited Pole Station. National operators were generally reluctant to establish mixed communities, often apparently assuming that small numbers of women in a male-dominated community would give rise to serious management problems. Change was therefore slow but by 1975 there had been women scientists wintering at McMurdo and female logistic staff. The United States continued to recruit regardless of gender and these days women make up around 30% of the people working at McMurdo in summer. Australia followed the American lead with a woman wintering as the Macquarie Island doctor in 1976 and in 1981 another woman doctor at Davis Station. Very slowly other countries introduced women into their programmes, initially as summer participants and usually as scientists or doctors. Nowadays, virtually all Antarctic countries have an open opportunity policy that is gender neutral, but the proportion of women on most stations is still fairly low as there are relatively few women in the physical sciences and even fewer technically qualified as plumbers, engineers, carpenters and other tradespeople. However, the mixed communities typical of present Antarctic stations have changed the ethos, making it easier to recruit highly qualified staff now that women are not excluded, providing a more normal atmosphere whilst not significantly changing the usual range of management problems.

Figure 8.27 Rose Meyer gets pretty cold when refuelling airplanes at Amundsen–Scott South Pole Station. (Credit: Kristina Scheerer, NSF)

equipment and understanding the science that will be taking place during the year. Everyone needs to be up-to-date in their general first aid certification and most go through a more intensive first aid training that is specific to the harsh environment and likely challenges they will face. Most programmes also take staff through fire-fighting training. Fire is an extreme danger in Antarctica where there is very little fresh water and the impact of fire on the station could easily render everyone homeless.

Programmes differ, but most require everyone who travels to the Ice to complete an Antarctic Field Training course, which may begin in the home country but then be followed by practical training immediately after arriving in Antarctica. For most scientists and short-term visitors, such as the media, these courses might last 1 or 2 days and cover the basics of survival in the extreme environment with practice in first aid, construction of emergency shelters and cooking and camping outdoors. For staff staying longer in Antarctica, though, the field training is typically more intense and for those on search and rescue teams, an even more detailed programme is followed. Whilst safety records around the continent are impressive (there are fewer than 10 field incidents per year, ranging from the most minor to major crevasse rescues), the search and rescue teams must be prepared for the possibility of catastrophic disasters.

The challenges of working in Antarctica have not gone away, although experience, planning, good training and modern technology have lessened the risks for most people. There are still the problems inherent in living and working in small communities, and in winter the rest of the world does seem a very long way over the horizon, despite the email and telephone connections. Working there will always be an experience for the few rather than the many.

Through all the complexity of life in Antarctica runs the spirit of international cooperation. Having a human presence there is costly and fraught with risk. However, it is through the collaboration of national programmes that the principles of peace and science, central to the Antarctic Treaty, can be realised. Perhaps it is the example Antarctica presents to the world that means the experience of the place stays with those who have been there long after they are gone.

9 | Scientists together in the cold

COLIN P. SUMMERHAYES

> The Antarctic is proof that great things can be achieved through
> collaboration. **Richard Fifield**

Early steps in Antarctic collaboration

Scientific work has always been conducted at a range of scales from the
individual to enormous international consortia, where the equipment needed
or the questions being asked are bigger than any one person or nation can manage.
This is particularly true for those scientific questions that cross geographic
boundaries or are at the scale of whole ocean basins, whole continents or of the
planet itself and its relation to other planetary bodies. Antarctica is one of these
large-scale challenges for science.

The International Geophysical Year (IGY) of 1957–58 brought together many
thousands of men and women from all over the world to take part in a common
undertaking on a scale never before attempted in science. It was organised by
scientists, yet supported by governments. It was a civilian exercise, yet logistically
supported in many countries by the armed forces. It was watched over and
coordinated, but not organised by, an international institution, yet relied on the
collaboration of scientists from 67 countries. Despite its apolitical nature, its
cooperative spirit contributed in no small way to the diplomatic framework for
later negotiations leading to such developments as the Antarctic Treaty (1961),
the Test Ban Treaty (1963) and the Space Treaty (1967).

As well as all this, the IGY also captured the imagination of the public and
politicians alike, demonstrating how much more we needed to know about our own
planet. Perhaps the greatest of its impacts came from the launch by the Soviet Union

Antarctica: Global Science from a Frozen Continent, ed. David W. H. Walton. Published by Cambridge University
Press. © Cambridge University Press 2013.

of the first satellite, Sputnik-1, on 4 October 1957 – which stimulated the space race and the US Moon programme.

During the IGY, the International Council of Scientific Unions (ICSU – now the International Council for Science) recognised the need to continue Antarctic collaborative research beyond the end of the IGY observing period. Three so-called 'Special Committees' were created to support this, and grew into three Interdisciplinary Bodies of ICSU – the Scientific Committee on Oceanic Research (SCOR), the Committee on Space Research (COSPAR) and the Scientific Committee on Antarctic Research (SCAR).

There was, of course, some international coordination in the Antarctic prior to the IGY, starting with the first International Polar Year (IPY) of 1882–83, when there was a French observing station on Tierra del Fuego, near Cape Horn, and a German one on South Georgia. The several individual national expeditions to Antarctica itself at the end of the nineteenth and beginning of the twentieth century during the so-called 'Heroic Age of Exploration' could also be seen as part of an international endeavour, having been stimulated by the International Geographical Congress of 1895, which made Antarctica the main target for new exploration. The second IPY offered an opportunity to expand upon the first, but coming as it did in 1932–33, at the height of the global economic depression, efforts were focussed in the Arctic; those in the south were restricted to observations from stations in South Georgia and the South Orkneys.

These various early Antarctic activities left no permanent establishments and did not form the start of any permanent activity centred on Antarctica in the way that the IGY did. Nor was science ready at that point for international collaboration of the kind or on the scale that we accept as normal today.

The Scientific Committee on Antarctic Research (SCAR)

SCAR was born from the unanswered questions and the exciting new horizons that the IGY opened up in the Antarctic. Charged with continuing the coordination of scientific activity in the Antarctic that had begun during the IGY, and with developing a more extensive scientific programme of circumpolar scope and significance, SCAR's first meeting, in The Hague on 3–5 February 1958, set out to build on the enthusiasm and the opportunities that the new Antarctic infrastructure could make possible. From the very beginning, the SCAR mission has been to facilitate and coordinate Antarctic research at a pan-Antarctic level beyond that possible by its individual national members.

What this means in practice is that SCAR helps the Antarctic scientific community to initiate, develop, and coordinate high quality international scientific research on and around Antarctica, as well as investigating the role of Antarctica and

Figure 9.1 The First Meeting of SCAR, The Hague, February 1958: 1. Dr L. M. Gould, United States; 2. Dr Ronald Fraser, ICSU; 3. Dr N. Herlofson, Convenor; 4. Col. E. Herbays, ICSU; 5. Professor T. Rikitake, Japan; 6. Professor Leiv Harang, Norway; 7. D. Valter Schytt, IGU; 8. Dr Anton F. Bruun, IUBS; 9. Mr J. J. Taljaard, South Africa; 10. Capt. F. Bastin, Belgium; 11. Capt. Luis de la Canal, Argentina; 12. Sir James Wordie, UK; 13. Professor K. E. Bullen, Australia; 14. Dr H. Wexler, United States; 15. Ing. Gén. Georges Laclavère, IUGG; 16. Ing. Gén. André Gougenheim, France; 17. Mr Luis Renard, Chile; 18. Dr M. M. Somov, USSR; 19. Professor J. van Mieghen, Belgium.

its surrounding Southern Ocean in the Earth System. To make this pan-Antarctic approach work, SCAR relies on free and unrestricted access to Antarctic scientific data and information from all Antarctic nations, a feature that is supposed to be guaranteed by the Antarctic Treaty, but which works less well in practice than was originally hoped for.

SCAR's national delegates meet every two years to set priorities and discuss progress. Actual day-to-day organisation is the responsibility of a small SCAR Secretariat, based at the Scott Polar Research Institute in Cambridge, UK.

Back in 1958 (Table 9.1), SCAR had just 12 member countries, and four ICSU scientific unions (IGU, covering geography and glaciology, IUBS for biology, IUGG for geodesy and geophysics and URSI for radio science – meaning all aspects of

Table 9.1 SCAR's National Members in 2010 (* = original 12 members marked; Russia was then the USSR, which included Ukraine)

Argentina *	Australia *	Belgium *	Brazil	Bulgaria
Canada	Chile *	China	Denmark	Ecuador
Finland	France *	Germany	India	Italy
Japan *	Korea	Malaysia	Monaco	Netherlands
New Zealand *	Norway *	Pakistan	Peru	Poland
Portugal	Romania	Russia *	South Africa *	Spain
Sweden	Switzerland	Ukraine *	UK *	USA *
Uruguay				

electromagnetic fields and waves). Things stayed that way for 20 years until Poland and the Federal Republic of Germany joined in 1978. By 2010 SCAR had 36 member countries and counted among its members nine of ICSU's Scientific Unions, the additional five being IUGS for geological sciences, IUPAC for chemistry, IUPS for physiological sciences, INQUA for Quaternary research, and the International Astronomical Union (IAU).

With minimal staff and little money SCAR's work can only be carried out through unpaid groups of like-minded scientists who see how collaborations can open up new horizons for them, both intellectually and in access to infrastructure.

Since its reorganisation in the year 2000, SCAR has focussed its efforts on a limited number of major scientific research programmes that address significant topical issues over a 5–10 year period (Table 9.2). They are intended to make significant advances in our understanding of how the Antarctic region works, and its role in the global system – in particular: documenting past change; detecting present change; evaluating the environmental effects of change; attributing causes; and improving the ability to forecast future trends. SCAR provides these major programmes with limited funds to support the meetings and workshops needed for them to succeed. In addition, but with fewer resources so as not to detract from the overall focus, it supports other types of groups to work on specific scientific topics chosen by the scientific community: *Action Groups* address specific matters with a narrow remit and normally last only 2–4 years. *Expert Groups* address matters on a longer time-scale. These various activities are managed by one or other of three Standing Scientific Groups (SSGs) the members of which are appointed by national academies; these are the SSGs on Life Sciences (SSG-LS), on Physical Sciences (SSG-PS) and on the Geosciences (SSG-GS). Table 9.2 represents, therefore, a

Figure 9.2 Five of the first six Presidents of SCAR. Left to right: George Knox (New Zealand, 1978–82); Torre Gjelsvik (Norway, 1974–78); Jim Zumberge (United States, 1982–86); Georges Laclavère (France, 1958–63); Gordon Robin (United Kingdom, 1970–74); Larry Gould (United States, 1963–70) is absent.

current snapshot of what is a rolling programme, some elements of which will change at every biennial SCAR meeting. Details of most of SCAR's programmes, including results and reports, are available on the SCAR website at www.scar.org.

SCAR also has committees of specialists with interests in data and information management, in Antarctic geographic information, and in the provision of advice to the Antarctic Treaty System, as well as a small group dealing with the history of Antarctic research.

SCAR's seed corn funding assists all the groups in the development of activities, such as workshops and websites, especially where there could be concrete outcomes, such as scientific papers in journals, meeting reports, databases, maps and other products. Less tangible but no less valuable outputs are the recognition of international excellence through the awards of fellowships to young researchers and the medals for outstanding scientific achievements and international cooperation.

Whilst supporting collaboration between individuals SCAR itself works in partnership with the national operators, who are responsible for logistics and who work together through the Council of Managers of National Antarctic Programmes (COMNAP). SCAR also works in partnership with global scientific organisations with a regional interest in the Antarctic, such as the World Climate Research Programme (WCRP). Since 2004 SCAR has also partnered its Arctic counterpart, the International Arctic Science Committee (IASC), in several areas of bipolar interest, especially climate change. Through these links SCAR is currently developing plans for the observing systems that will capture the knowledge of oceanic and ice systems in the Antarctic that we need as the basis for improved forecasts of change.

Not all member countries have large and experienced Antarctic communities, and so there is a need to share expertise and access to specialised equipment to improve national capabilities. For young scientists, the SCAR Fellowship Programme, launched in 2003, provides an opportunity to work in another country whilst the biennial Open Science Conferences, launched in 2004, provide a regular major

Table 9.2 **The SCAR Science Groups in 2010**

Type	Title
Scientific research programmes	Antarctic Climate Evolution (ACE)
	Subglacial Antarctic Lake Environments (SALE)
	Antarctica and the Global Climate System (AGCS)
	Astronomy and Astrophysics from Antarctica (AAA)
	Evolution and Biodiversity in the Antarctic (EBA)
Action groups	Sub-Ice Geological Exploration (SIeGE)
	Acoustics in the Marine Environment
	Census of Antarctic Marine Life (CAML)
	Marine Biodiversity Information Network (MarBIN)
	Antarctic Fuel Spills (AGAFS)
	Seeps and Vents Antarctica (SAVAnt)
	Environmental Contamination in Antarctica (ECA)
	King George Island Science Coordination Group
	GPS for Weather and Space Forecasting
	Polar Atmospheric Chemistry at the Tropopause (PACT)
	Prediction of Changes in the Physical and Biological Environment of the Antarctic
Expert groups	Geospatial Information – Geodesy (GIANT)
	Permafrost and Periglacial Environments (PPE)
	International Bathymetric Chart of the Southern Ocean (IBCSO)
	Antarctic Digital Magnetic Anomaly Project (ADMAP)
	Continuous Plankton Recorder Research (CPR)
	Birds and Marine Mammals (EGBAMM)
	Human Biology and Medicine
	International Partnership in Ice Core Sciences (IPICS)
	Oceanography Expert Group [joint with SCOR]
	Operational Meteorology in the Antarctic (OpMet)
	Ice Sheet Mass Balance and Sea Level (ISMASS)
	Inter-hemispheric Conjugacy Effects in Solar-Terrestrial and Aeronomy Research (ICESTAR)
	Antarctic Sea Ice Processes and Climate (ASPeCt)
	International Trans-Antarctic Expedition (ITASE)

Figure 9.3 Flyer from XXIX SCAR Conference, Hobart, 2006. (Credit: Ian Allison)

forum for planning and exchange and now attract over 1000 scientists. SCAR is also keen to help promote the incorporation of Antarctic science in education at all levels, and to communicate scientific information about the Antarctic region to the public, notably by means of the internet.

Key SCAR scientific achievements

SCAR programmes have made major contributions to understanding such diverse phenomena as: Antarctic weather and climate; the evolution of Antarctic climate; changes in the volume of the Antarctic ice sheet; the distribution of sea ice; the distributions of seabirds and seals; invasive species; the suitability of Antarctica for an astronomical observatory; an understanding of Sun–Earth interactions in the upper atmosphere; subglacial lake environments; geodesy; magnetism; and the biodiversity of the Antarctic flora and fauna.

Key achievements in recent years include the following:

- Determining the functional ecosystem processes of the Southern Ocean ecosystem, including the key role of krill (El-Sayed, 1994), and documenting the distribution, abundance and long-term trends in Antarctic and sub-Antarctic seabirds.
- Understanding the diversity, ecology and population dynamics of the organisms beneath the Antarctic sea ice, and their sensitivity to change.
- Establishing how Antarctic land, lake and pond life respond to climate change, and identifying the processes determining community response to stress (Bergstrom *et al.*, 2006).
- Discovering a major warming of the Antarctic winter troposphere, 5 km above sea level, that is larger than any other tropospheric warming on Earth.
- Confirming that, while the Antarctic Peninsula has warmed significantly (3°C on average and 5°C in winter on the west coast over the past 50 years), air temperatures in East Antarctica have remained steady or cooled.
- Determining that the Larsen-B Ice Shelf collapsed because prevailing westerly winds brought more warm air across the Antarctic Peninsula as the planet warmed.
- Providing the basis for determining the mass balance of the Antarctic Ice Sheet. West Antarctica is losing mass; East Antarctica remains largely stable.
- Developing a climatology of Antarctic sea ice for understanding sea ice formation, validating satellite data and feeding coupled ocean–ice–atmosphere models.
- Generating plans and guiding principles for the exploration and environmental stewardship of unique, pristine, subglacial lakes.
- Developing and publishing plans for a cryosphere observing system.
- Developing and publishing plans for a Southern Ocean Observing System.

Some of these achievements have come about through the development of major scientific programmes, such as the Biological Investigations of Marine Antarctic Systems and Stocks (BIOMASS) Programme (1977–91), which SCAR led in a consortium including the Scientific Committee on Oceanic Research (SCOR), the International Association of Biological Oceanography (IABO), and the Food and Agriculture Organisation (FAO) of the United Nations. Ships from 12 countries

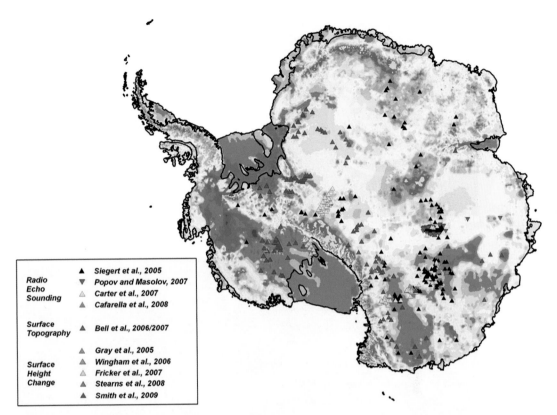

Figure 9.4 The SCAR SALE programme map of 145 subglacial lakes. (Credit: Martin Siegert)

made 31 cruises spread over the summer seasons of 1980–81, 1983–84 and 1984–85 to establish the importance of krill in the Southern Ocean food web and to estimate how much krill was available before much fishing took place. As with all SCAR programmes, the BIOMASS fieldwork and analysis were funded through national programmes. SCAR provided coordination in planning operations and integrating results. BIOMASS was important not only for establishing the role of krill in the Southern Ocean ecosystem, but also for leading directly to the creation of the Convention on the Conservation of Antarctic Marine Living Resources (CCAMLR), and its adoption of an ecosystem-based approach to fisheries management. The BIOMASS data initiated the CCAMLR database. The modern equivalent of BIOMASS is SCAR's Census of Antarctic Marine Life (CAML) programme, a major international effort involving some 17 ships, which had its first cruise in December 2006 and which ended in 2010.

International recognition of SCAR's role as a forum for coordination and collaboration was provided by the award of the prestigious Prince of Asturias' prize in 2002 for international cooperation. The prize money of 50 000 euros enabled SCAR to begin its fellowship programme for young scientists.

Figure 9.5 The BIOMASS logo.

Figure 9.6 The award of the Prince of Asturias Prize to SCAR for international collaboration. Left to right: Peter Clarkson, Jeronimo Lopez-Martinez, Robert Rutford and Roland Schilch. (Credit: Jeronimo Lopez-Martinez)

Figure 9.7 An illustration of a bedrock topography of Antarctica from a map sheet compiled by SCAR from international data.

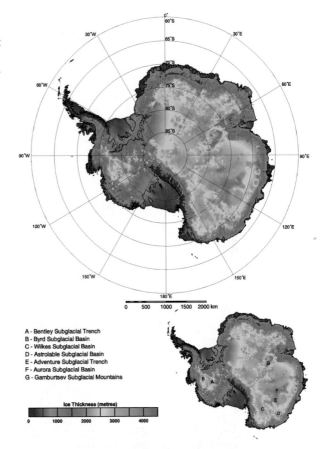

A - Bentley Subglacial Trench
B - Byrd Subglacial Basin
C - Wilkes Subglacial Basin
D - Astrolabe Subglacial Basin
E - Adventure Subglacial Trench
F - Aurora Subglacial Basin
G - Gamburtsev Subglacial Mountains

SCAR products

There is no single centralised database for Antarctica. Instead, SCAR, recognising that there is a wide range of data and database systems in its member countries, has organised a meta-data directory the Antarctic Master Directory (AMD), to guide people to much of the data that has been collected. This is a web-based metadata catalogue populated with contributions from National Antarctic Data Centres (http://gcmd.gsfc.nasa.gov/KeywordSearch/Home.do? Portal=amd&MetadataType=0). National Antarctic Data Centres (NADCs) or their equivalents have now been established in most SCAR countries, and – together with the AMD – comprise the Antarctic Data Directory System (ADDS). Growing steadily, the ADDS is intended eventually to provide an index for access to all Antarctic data, no matter where or how they are stored. SCAR recently published a SCAR Data and Information Management Strategy designed by SCADM to help all Treaty nations improve the way data and information is managed and exchanged.

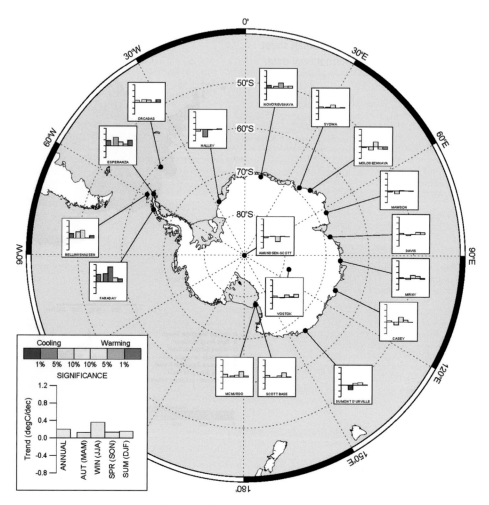

Figure 9.8 Estimates of Antarctic near-surface temperature trends for 1951–2006 based on station data. (Credit: John Turner, BAS).

SCAR also supports the activities of the scientific community through the development of a range of additional essential products (Box 9.1). These are available on line to the scientific community, the national science and logistics operators, the Treaty Parties and the public.

Science and policy

SCAR is funded by national academies of science affiliated to ICSU. Because it is not an intergovernmental body, SCAR's agenda is nominally independent of national aims. Nevertheless, much of what SCAR does in practice is facilitated

Box 9.1 **SCAR Products (available from www.scar.org)**

The REference Antarctic Data for Environmental Research (READER) project has created databases for meteorological data, ice core data and ocean data.

The Antarctic Digital Database (ADD) provides topographic data for Antarctica, and provides access to maps of Specially Protected Areas, Historic Sites and Monuments, and Seal Reserves.

SCAR maintains a web-accessible Antarctic Biodiversity Database whose contents contribute to the Global Biodiversity Information Facility (GBIF), and the Ocean Biodiversity Information System (OBIS). SCAR's Marine Biodiversity Information Network (MarBIN) compiles and manages information on Antarctic marine biodiversity through a distributed system of interoperable databases.

The Composite Gazetteer of Antarctica is a comprehensive compilation of the various national lists of place names in Antarctica.

The Seismic Data Library System (SDLS) makes Compact Disc copies of seismic data over 4 years old available for joint projects and distributes them to regional libraries.

Those interested in geodesy can consult SCAR's Master index for Antarctic positional control, and SCAR's Geodectic Control Database, which provides access to high precision positional data from seven countries, and which is useful for aerial photography, mapping and satellite imaging projects.

There is an Antarctic Map Catalogue, which contains information on all maps published by SCAR Members.

BEDMAP comprises Antarctic Bedrock Mapping Data collected on surveys undertaken over the past 50 years, and describing the thickness of the Antarctic ice sheet.

Tide gauge data on sea level measured around Antarctic are managed by the Permanent Service for Mean Sea-Level (PSMSL).

The IBCSO programme is producing a new bathymetric chart of the Southern Ocean.

The Antarctic Digital Magnetic Anomaly Project (ADMAP) maintains the magnetic anomaly database of Antarctica.

A Geographical Information System for King George Island (KGIS) facilitates the work of the several SCAR Members who maintain national bases there.

The SCAR Feature Catalogue will enable geographic database interoperability in the Antarctic community.

The Continuous Plankton Recorder programme compiles data from all oceanographic cruises in the Southern Ocean that collect plankton samples from a towed Continuous Plankton Recorder device.

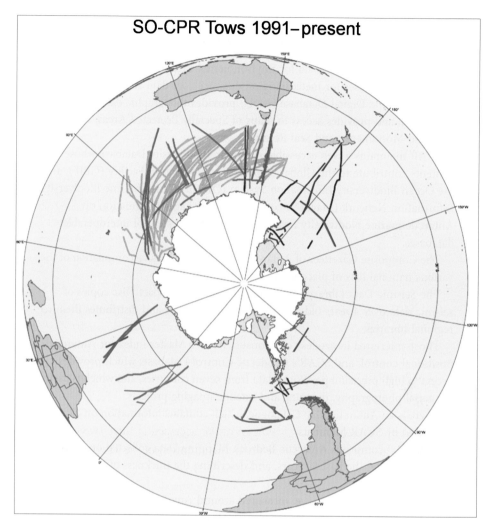

Figure 9.9 Continuous Plankton Recorder transects in the Southern Ocean by several nations as part of a SCAR coordinated programme. (Credit: Graham Hosie, Australian Antarctic Division)

by national Antarctic logistic operators, and both uses and adds value to the activities of national programmes. Indeed, many of SCAR's scientists may be government scientists working alongside those from academia. Starting in 1961, SCAR was invited to provide advice to the newly formed Antarctic Treaty, which was another product of the IGY. Since the first Antarctic Treaty Consultative Meeting, SCAR has continually provided Antarctic Treaty Parties with objective and independent scientific advice on issues of science and conservation affecting the management of Antarctica and the Southern Ocean. More recently, it has passed

Figure 9.10 Anna Jones (UK), seated by the open laptop computer at left, delivers the first SCAR Lecture to the Plenary at XXVI ATCM in Madrid in 2003. (Credit: Peter Clarkson, SCAR).

such advice also to younger related organisations, such as CCAMLR. SCAR papers presented to the annual Treaty meetings can be accessed at http://www.scar.org/treaty/. SCAR also brings topical science to the attention of the Treaty Parties via the SCAR lecture at each Antarctic Treaty Consultative Meeting (ATCM). A significant recent accomplishment was publication of the Antarctic Climate Change and the Environment Review (ACCE) (www.scar.org/publications/occasionals/acce.html).

Through its advice, SCAR has significantly influenced the way the Antarctic Treaty has developed since being signed into law in 1961 (Table 9.3), but walking the line between independent advice and politically charged agendas has never been easy.

SCAR and the Fourth International Polar Year (2007–09)

This is not the place for a comprehensive review of the recent IPY. Suffice it to say that the fundamental concept of the IPY was of an intensive burst of internationally coordinated, interdisciplinary, scientific research and observations focussed on the Earth's polar regions during an official observing period from 1 March 2007 until 1 March 2009. The IPY would aim to exploit the intellectual resources and science assets of nations worldwide to make major advances in polar knowledge and understanding, while leaving a legacy of new or enhanced

Table 9.3 Contributions from SCAR to the development of Antarctic Treaty Instruments

1	Provided advice that led in 1964 to the adoption by the Antarctic Treaty of the Agreed Measures for the Conservation of Flora and Fauna
2	Provided further advice leading in 1991 to the Agreed Measures forming the core of a more comprehensive environmental agreement – the Protocol of Environmental Protection to the Antarctic Treaty
3	Developed the original concepts of Sites of Special Scientific Interest and Specially Protected Areas for Antarctica
4	Provided exemplar framework for management plans for Antarctic Specially Protected Areas (ASPAs) based on Moe Island
5	Provided a Management Plan Handbook and a Visit Report Form, as well as the scientific advice to modify and edit plans for these sites
6	Designed the Checklist for Environmental Inspections under the Antarctic Treaty
7	Together with COMNAP, developed the Environmental Impact Assessment (EIA) Guidelines and good practice
8	Together with COMNAP, developed the Environmental Monitoring Handbooks
9	Organised the workshop with IUCN that put environmental education onto the ATCM agenda
10	Provided key advice that led to the Treaty Parties adopting the IUCN criteria for listing and delisting species
11	Provided the advice that led to the delisting of Fur Seals
12	Through BIOMASS, provided the foundation for the creation of the Scientific Committee of CCAMLR
13	BIOMASS database was adopted by CCAMLR as the basis for its initial work programme
14	Provided CCAMLR with data on higher predators
15	Heavily involved in initiating and developing the Convention for the Conservation of Antarctic Seals
16	Published reports containing advice for the negotiation of the Convention on the Regulation of Antarctic Mineral Resource Activities (CRAMRA)
17	Developed Codes of Conduct for (a) Fieldwork, and (b) the Use of Animals for Scientific Purposes in Antarctica

observational systems, facilities and infrastructure. The most important legacies should be a new generation of polar scientists and engineers, as well as an exceptional level of interest and participation from polar residents, schoolchildren, the general public and decision-makers worldwide. The IPY would strengthen international coordination of research and enhance international collaboration and cooperation in polar regions. Interdisciplinary approaches would be emphasised to address questions and issues lying beyond the scope of individual disciplines. The timing was thought to be especially appropriate given the rapidly accumulating signs that global warming was having its most powerful effect in the polar regions, where there was ample evidence for shrinking sea ice, melting permafrost and retreating glaciers.

SCAR played a major role in stimulating the international consensus for the IPY and helping to organise it. The idea of celebrating the 50th anniversary of the IGY and the 125th anniversary of the first IPY arose in 1999. In 2002 SCAR Delegates supported the proposal and suggested that enquiries should be made to ICSU and IUGG. SCAR Vice-President Professor C. G. Rapley agreed to follow up this proposal. After the US National Academy of Sciences confirmed its feasibility through an international workshop, Robin Bell (United States) and Chris Rapley (UK) jointly presented ICSU with a proposal. Following the report of the ICSU Planning Group in 2004, ICSU agreed that WMO should become a co-sponsor in organising the IPY. The two organisations formed a Joint Committee to steer the process, and an international programme office was established at the British Antarctic Survey in Cambridge to manage the process.

The IPY projects can be reviewed on the IPY web site at www.ipy.org. Not only were most endorsed IPY projects strongly internationally collaborative and interdisciplinary, they were also cross-thematic, most being targeted at more than one of the IPY science themes. The international nature of the individual projects, and their common interdisciplinary nature, were further departures from the IGY and previous Polar Years, where single discipline science projects tended to be effected by single nations. This broad international effort should contribute to a future of increased cooperation between scientists, organisations and nations in the knowledge and rational use of our planet.

SCAR began contributing as an observer to the meetings of the IPY Planning Committee, starting in March 2004, and encouraged its scientific community to contribute proposals to the IPY steering committee. As a result, all of SCAR's large Scientific Research Programmes, and several of its Action and Expert Groups ended up leading or being much involved in programmes approved by the IPY Joint Committee.

SCAR played a key role in the IPY, wielding appropriate influence in the way in which the IPY was steered. Recognising the key role of SCAR in the IPY, the Joint Committee decided that the 2008 SCAR/IASC Open Science Conference in St Petersburg would be the first of three major IPY Conferences. The second in the series addressed progress at the end of the IPY, and took place in Oslo, Norway, from 8–10 June 2010. The third will review the implications of science for policy makers, and will take place in Montreal, Canada, in 2012. SCAR expects to take on a

Figure 9.11 The Executives of SCAR and IASC at the first joint Open Science meeting in St Petersburg in 2008, which was also the first of the three IPY Science Conferences. From left to right: Elena Manaenkova (WMO), Artur Chilingarov (Russian Duma), Louwrens Hacquebord (IASC Co-chair of the Organising Committee), Vladimir Kotlyakov (Co-chair of the Local Organizing Committee), Khotso Mokhele (ICSU), Chuck Kennicutt (SCAR Co-chair of the Organising Committee), José Retamales (Chair of COMNAP), Chris Rapley (SCAR President) and Kristjan Kristjansson (IASC President).

major role in managing the Antarctic legacy of the IPY, in much the same way that it has successfully managed the Antarctic legacy of the IGY.

The international importance of Antarctica is now clearly recognised by many governments and its science is playing a major role in validating the

Figure 9.12 Celebratory cake to commemorate the first joint SCAR/IASC meeting in St Petersburg. (Credit: Colin Summerhayes)

Figure 9.13 The SCAR flag flying at Neumayer Station, with SCAR Executive Director Colin Summerhayes, November 2004 (Credit: Peter Clarkson)

models for climate prediction. The role of the Southern Ocean not only as a carbon sink but also as a centre for evolution of many deep-sea organisms and as the locus of a sustainable fishery are also of great significance whilst the search for the secrets of genetic adaptation to the cold and the production of secondary compounds by many marine organisms is fuelling possible commercial interests from pharmaceutical companies. The future of world sea level is intimately tied to the Antarctic ice sheet whilst the ice itself provides us with a way of monitoring the way in which we are polluting our world. In all these many ways and more Antarctic science has an important role to play in our global future, one which SCAR will continue to help and develop.

10 Managing the frozen commons

OLAV ORHEIM

> The Antarctic Treaty is indispensable to the world of science which knows no national or other political boundaries, but it is much more than that... it is a document unique in history which may take its place alongside the Magna Carta and other great symbols of man's quest for enlightenment and order. **Laurence Gould, 1960**

A lawless century: the early years of sealing and whaling

When James Cook had circumnavigated Antarctica in 1775 he reported an abundance of seals, and this caught the attention of the sealing community in the north. Coming mostly from the United States and from the United Kingdom, they hunted both the fur seals for their fur, and the elephant seals for oil. Sealing continued for over a century, but with varying economic returns. The activities peaked around 1820. It has been estimated that a third of the estimated 5.2 million fur seals killed in the south were killed within the Antarctic region. The sealing period was one of no regulations in the southern ocean, and the sealers did not in general publish their geographic discoveries, as such knowledge was of commercial value.

However, several national expeditions carried out geographic explorations, and their activities are well documented. The expedition led by Fabian Gottlieb von Bellingshausen of the Russian Imperial Navy was probably the first to see Antarctica on 27 January 1820. This was 3 days before Edward Bransfield of the British Royal Navy recorded sighting land. Almost a year later, the American sealer Nathaniel Palmer also reported having seen land, and his countryman, the

Antarctica: Global Science from a Frozen Continent, ed. David W. H. Walton. Published by Cambridge University Press. © Cambridge University Press 2013.

THE ANTARCTIC SEA-ELEPHANT FISHERY.

Stripping blubber and rolling it in barrels to try-works, Southwest Beach, Herd's Island.　(Sect. v, vol. ii, pp. 419, 435.)

Drawing by H. W. Elliott, after Capt. H. C. Chester.

Figure 10.1 Stripping sea elephant blubber and rolling it in barrels to try-works, Southwest beach, Herd's Island. From G. B. Goode (1887). *The Fisheries and Fishery Industries of the United States*. (Credit: NOAA)

sealer John Davis, may have made the first landing on mainland Antarctica on 7 February 1821. These expeditions would later be used as building blocks in territorial discussions.

It was the introduction of large-scale whaling, a much more important commercial activity, which brought forward the need for regulations. As with the sealing, it was technology from the northern hemisphere that made the impact.

In 1892–93 four expeditions went south, three Norwegian and one Scottish. C. A. Larsen led the *Jason* expeditions, which shot the first whale in the Southern Ocean. These expeditions had financial backing from magnates of whaling in northern waters. Although the initial expeditions were not a financial success, the observations of multitudes of large whales led to several companies becoming involved, primarily from Norway. By the early twentieth century, such companies were establishing whaling stations on South Georgia and on Deception Island.

Figure 10.2 Grytviken Whaling Station on South Georgia. Established in 1904 and abandoned in 1965 this photo shows it in its original state before buildings were removed to make it safe. (Credit: Peter Bucktrout , BAS)

The sectorial claims and the problems between the UK, Argentina and Chile

In the nineteenth century the strongest European nations had established colonies by force in many parts of the world. They exerted control and applied their own laws and regulations. Gradually exploration proceeded also into the inhospitable polar regions. When it came to Antarctica, where there were no inhabitants, the accepted practice for establishing a claim was to be the first to set foot on land, and take possession by describing and mapping it. Naming was crucial – this right fell to those who first laid eyes on the landscape. Annexation began with the expedition leader standing on the land in question, solemnly claiming it in the name of his nation and erecting a sign to mark the action. It was always 'his', as only men led these expeditions!

For the claim to be recognised internationally, the claimant nation had to authorise the annexation, and there had to be no other nations protesting against it. If there were formal protests, these had to be dealt with. This could occur through negotiations or the use of force; sometimes protests were simply ignored.

In these early days it was British expeditions that led the exploration of the continent, although there were also notable expeditions from Australia, Belgium, France, Germany, Japan, Norway and Sweden.

It was the Norwegian whaling companies that speeded up annexation. They wanted to establish facilities on land and asked the British government for permits to do this. In 1908 the United Kingdom established the Falkland Islands Dependencies, covering the sector from 20°W to 80°W (later renamed the British Antarctic Territory). At the same time it put in place licensing fees for whales caught in this sector, and allowed establishment of whaling stations on South Georgia and in the South Shetland Islands.

At this time, the British had the world's strongest navy, and the claim was not challenged by force. But Argentine and Chilean interests were also present, through whaling activities in South Georgia and Deception Island, respectively. Argentina had also 4 years earlier taken over the meteorological station that Bruce's *Scotia* expedition had established on Laurie Island in the South Orkney Islands. This station, later renamed Orcadas, has been in operation ever since and is the station in Antarctica with the longest records.

The next sector to be claimed was also a result of British and Commonwealth explorations, namely New Zealand's claim from 1923 on the Ross Dependency, from 150°W to 160°E. Only a year later, France claimed Adélie Land, from 142°2′E to 136°11′E. In 1933 Australia claimed the Australian Antarctic Territory, which extends from 160°E to 45°E, except for the sector claimed by France. In 1939 Norway claimed Dronning Maud Land, from 45°E to 20°W. This is the only claim that is not a sector, as Norway does not adhere to the sector principle in the Antarctic. Norway also claimed Peter I Island, at 68°50′S 90°35′W, in 1929.

By 1939 the whole of the Arctic continent was claimed, except for a sector from 80°W to 150°W. The five claimant nations, who over the years have mutually recognised each other's claim, hoped that the United States would claim the remaining sector. The United States had been active in this area, and such a claim would have meant that all of Antarctica had been divided up. But this was not to be. Instead the United States, and also the Soviet Union/Russia, chose to maintain a claim on all of Antarctica, based on their explorations of the whole continent.

Formal jurisdictional disagreements were heightened when Chile in 1940 claimed a sector of Antarctica, from 53°W to 90°W, and when Argentina 3 years later did the same for the sector from 25°W to 74°W. Their positions were based more on contiguity than on primacy of presence. As can be seen the Argentine, British and Chilean claims all overlap from 53°W to 74°W, which corresponds to the longitudes of the Antarctic Peninsula. On both sides there are sectors disputed by only two of the countries, and the British and the Chilean claims also contain segments that they claim alone.

Antarctica never became a battlefield during World War II. But there were worries that the Antarctic Peninsula could be used to support German or Japanese maritime operations, so the British deployed small military units there in 1943, purporting to be watching for the enemy but this was only a cover story for strengthening a sovereignty claim. In the following years the conflicting claims between these three nations led regularly to formal protests, and also to some minor skirmishes between various stations in the Peninsula. But the harshness of surviving in Antarctica meant that the people stationed there on the whole developed ways of cooperation, rather than aggression.

Increasing international interest

After World War II, Antarctic exploration changed in scale and focus. This can be illustrated by two very different expeditions, which each in their way influenced international thinking about the continent.

The United States Navy conducted Operation Highjump in 1946, which remains by far the largest expedition ever to take place in Antarctica. It involved 13 naval ships including a US Coast Guard ice-breaker, 33 airplanes, a submarine and about 4800 men. For comparison, the approximately 30 countries now conducting activities in Antarctica bring altogether only 3–4000 persons to the continent each the summer. Unfortunately, the results of the expedition were not commensurate with the effort. The size of the operation can be partly attributed to the fact that there were many troops who had not yet been demobilised but it was also meant to demonstrate the interest of a large country in Antarctica. Its objectives included training men in polar climates and extending and consolidating US sovereignty over Antarctic areas. In the emerging Cold War this did not go unnoticed by the Soviet Union, which had not been active in the Antarctic, but had a much larger community of polar experts from its Arctic activities.

The other noteworthy Antarctic effort was totally different. This was the Norwegian–British–Swedish expedition of 1949–52, also named the Maudheim expedition after the name of its station. It was different from all previous expeditions in two main ways; it was truly international in science cooperation and in its financing, and its whole programme was based on testing scientific ideas. It paved the way for the science focus and the international cooperation that would be paramount 8 years later, when the International Geophysical Year (IGY) began, and the Scientific Committee on Antarctic Research (SCAR) was established.

The team at Maudheim consisted of very promising young scientists from several nations, and probably no other polar expedition has ever been such a springboard for eminent research careers. Two of these are especially important in the history of Antarctic science: Valter Schytt, who became professor at Stockholm University, was the first secretary of SCAR from 1958–59. He was

Figure 10.3 The Norwegian–British–Swedish expedition 1949–52 was the first planned international Antarctic research expedition. **(a)** Norwegian leaders by the station signpost. From left: Expedition leader John Giæver, director of the Norwegian Polar Institute, Harald Ulrik Sverdrup and captain of the expedition vessel Guttorm Jacobsen. (Credit: Hallgren); **(b)** Valter Schytt examining snow samples in microscope in the ice laboratory. (Credit: V. Schytt)

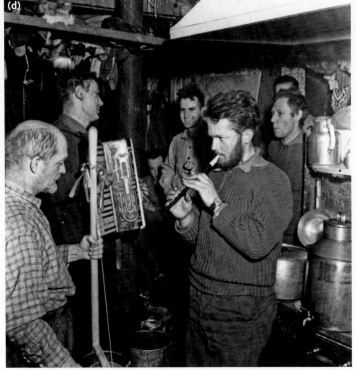

(c) Travelling to the mountains on a geophysical and surveying expedition. (Credit: H. Sverdrup); **(d)** Party time at Maudheim with Bjarne Lorentzen on the double bass, Egil Rogstad on the accordion and Gordon Robin on the flute. (Credit: All photos courtesy of the Norwegian Polar Institute)

immediately followed by Gordon de Q. Robin, who served in that capacity for 11 years, and then was president of SCAR in 1970–74. Robin was also director of the Scott Polar Research Institute in Cambridge, England, for 24 years, which has housed the SCAR office.

In the early 1950s it was agreed to have a new Polar Year in 1957–58 (see Chapter 1.) The International Geophysical Year (IGY) was approved to take place from 1 July 1957 to 31 December 1958. That the name used was 'Geophysical' and not 'Polar' reflected the broader approach compared to the previous two polar years. Particularly new was that 'space' was now on the agenda. Nevertheless, the main focus was again on the polar regions, and this time the largest effort was in Antarctica. About 50 year-round stations were established in the Antarctic, including the first stations far inland. Extensive research programmes were conducted.

The achievements of IGY, especially with respect to cooperation, were remarkable, given the political climate of the Cold War. In many ways IGY and its progeny, SCAR, became the catalysts for the political cooperation that was to follow, by facilitating and formalising the cooperation and exchange of scientists and data. So some brief comments on the early days of SCAR and Antarctic science are in order (see also Chapter 9).

The birth of the Antarctic Treaty

As IGY had a major Antarctic component, ICSU decided in September 1957 to set up a committee to organise this activity. Ten of the 12 nations engaged in Antarctic research, together with four international unions, met in February 1958 and elected the first officers for SCAR, and established working groups to prepare Antarctic research programmes after IGY. Since then all 12 nations have met regularly, and the number of SCAR members has now grown to 34.

The challenging Antarctic environment certainly helped establishing trust and understanding between the scientists and support personnel there. Occasionally accidents and other situations arose that were solved through joint international efforts. Scientists were exchanged, and a 'family' was formed of people with a common experience of overcoming the harsh climate. These personal ties were important when the political discussions started informally in June 1958.

No doubt another major reason for proceeding with such discussions was the attitude of the United States and the Soviet Union. By this time the Arctic Ocean had become a very important strategic arena in the military balance, especially for the Soviet Union, which had based its second strike capability on nuclear submarines carrying long-distance rockets. Neither country wanted an escalation into the Southern Ocean and Antarctica of another competitive military race.

Figure 10.4 The Plenary of the XXIX Antarctic Treaty Consultative Meeting in session in Edinburgh during the opening ceremony with Princess Anne. (Credit: Peter Bucktrout, BAS)

A third reason was that the three nations with competing claims on the Antarctic Peninsula also recognised the importance of avoiding the escalation of their conflict into military deployments to the Peninsula.

So when the United States in 1958 invited them all to discussions to consider developing a treaty for the continent the setting was apparently favourable. Clearly the negotiations were also simplified by the fact that there were no residents on the continent. The largest challenge that had to be tackled was how to resolve the questions of the conflicting Antarctic claims, including the positions of the superpowers that did not recognise the sectorial claims but instead maintained claims on the whole continent.

The negotiations were completed with the agreement on the Antarctic Treaty in Washington on 1 December 1959. It was signed by the 12 participating countries: Argentina, Australia, Belgium, Chile, France, Japan, New Zealand, Norway, South Africa, the Soviet Union (now Russia), the United Kingdom and the United States. These were the same countries as those that established SCAR almost 2 years earlier. The last of the 12 to ratify the Treaty did so on 23 June 1961, which is the date the Treaty entered into force.

The Treaty is remarkably simple in its form (see Box 10.1). It covers the area south of 60° S, with the expressed purpose of setting aside Antarctica for peace and science. It pertains to all land and ice shelves. The Treaty banned military activity on the continent other than for peaceful purposes. In this way the Treaty became the first arms control agreement established during the Cold War. To ensure that the agreement was adhered to the Treaty defined the right of international inspection. In making Antarctica a scientific preserve it also established formal requirements for the exchange of information.

The most fundamental provision is Article IV, which has 'frozen' the geographical claims in the pre-Treaty situation. While the Treaty is in force no new claims can be made, existing claims cannot be amended and no activities can affect the status of pre-Treaty claims.

The Treaty established biennial meetings, called the Antarctic Treaty Consultative Meetings (ATCM). Here the Contracting Parties would meet to consult and develop further the management of the continent. At the start the original 12 parties made up the so-called Antarctic Treaty Consultative Parties (ATCPs). At the ATCMs all decisions of substance need consensus from these parties. The Treaty opened for accession from other states. Such new members could participate fully in the discussions, but not in decisions.

The Antarctic Treaty has been a model for other international treaties, including the one for Outer Space and the inspection provisions in the ABM agreement between the Soviet Union and the United States.

In 2009 the 32nd ATCM was held, aptly enough, in Washington/Baltimore, 50 years after the signing of the Treaty.

The evolution of the Antarctic Treaty and its role in environmental management

The Treaty did not remove the disagreements on territorial claims; it suspended these, and removed the necessity of continually asserting the different national positions. By setting up a consultation mechanism based on consensus it established a remarkably successful system for developing solutions in managing the Antarctic. One might imagine that such a principle would lead to deadlocks over substantial issues. This has generally not been the case. Instead, the ATCPs have, over the years, developed a realistic understanding on how to work together even when there are fundamentally different positions. Understanding this practical application of consensus is critical to appreciate the evolution of the Treaty.

With the Treaty signed and ratified the 12 countries set about making it work. From before the first meeting of the Treaty Parties the Scientific Committee on Antarctic Research (SCAR) was directly involved. SCAR was a part of the International Council of Scientific Unions (ICSU) and thus a nongovernmental body divorced from politics. Consequently its close links with the Treaty has been a most unusual relationship, and yet one from which both sides have gained a great deal.

For more than a decade the consultative parties were only the original 12 nations, which were the same as for SCAR. It may be said with considerable justification that this was an 'Old Boys Club', as many of the participants at the meetings had known each other for years, and in some cases been in the Antarctic during the IGY. Today, the ATCPs are much less homogenous, and the membership of the Treaty and SCAR are not identical, although all Consultative Parties are members of SCAR.

The Parties realised immediately that they lacked the necessary scientific data on which to base important decisions about governance. The Treaty says very little about conservation and this needed to be addressed as a priority. SCAR drafted a proposal, which by 1964 had turned into the 'Agreed Measures for the Conservation of Antarctic Flora and Fauna', the first conservation initiative under the Treaty. And this became the practical way forward for the next few decades where SCAR sometimes initiated a new discussion whilst at other times the Parties asked formally for advice on particular subjects.

The Measures in 1964 were the first major step in a process that, over the years, has led to a whole complex of arrangements commonly called the Antarctic Treaty System (ATS). This consists of the Treaty with its many hundreds of Measures agreed to at the ATCMs, which thereafter have to be adopted by the individual parties to enter into force and become legally binding. In addition the ATS includes two separate conventions, on seals and marine life, and a protocol on Environmental Protection under the Treaty, all described below.

The ATCMs were originally every 2 years, but have been annual from 1991. Since 1983 the non-consultative parties have participated in the ATCM, and also in Special Consultative Meetings to consider issues of minerals, and of environmental protection. The Consultative Parties also invited observers and experts from intergovernmental and nongovernmental organisations to advise within their respective interests. From the start, SCAR has had a special observer status at the ATCMs, which later also has been given to COMNAP (Council of Managers of National Antarctic Programs), which was established in 1989. The two organisations give reports at the ATCMs. In this way the international community of the 'governed' – scientists and logisticians – have an independent route into the governance system. The ATCPs have also organised Meetings of Experts to consider special questions, such as telecommunications, air safety, tourism, code of shipping and climate change.

The responsibility for organising the ATCMs goes in rotation between the ATCPs, following the alphabet of country names in English. Until 2004 there was no permanent secretariat and no official archive. The nearest to the latter was the

Antarctic Handbook, which for years was updated by a dedicated delegate. So it stands to reason that this system was not easy for new nations to use, even if the old-timers did their best to help them along.

By July 2012, an additional 16 nations had attained consultative status, starting with Poland in 1977, and an additional 22 had acceded to the Treaty as non-consultative parties, the latest being Pakistan. So now the total number of Parties to the Treaty is 49. It is noticeable that around two-thirds of the global population lives in the nations that have signed up to the Antarctic Treaty. Large countries such as Brazil, China and India as well as many European and South American nations became consultative parties during the 1980s. At present the most populous countries outside the Treaty are Indonesia and Nigeria. Malaysia acceded to the Treaty as the first Muslim country in 2011, but with only one African country and no Arab states the membership is still not representative of the world's cultural diversity.

To become an ATCP a country has to fulfil Article IX, which states that an acceding country is a consultative party 'during such times as that Contracting Party demonstrates its interest in Antarctica by conducting substantial research activity there, such as the establishment of a scientific station or the dispatch of a scientific expedition'. The reasoning behind this was that the Antarctic region is such an exceptional part of the globe that to participate in decisions on its management, a party should be directly experienced in operations there through its own activities and able to demonstrate that it can make a substantive addition to scientific knowledge. However, there is no doubt that the interpretation of how to apply article IX has changed over the years. In the 1970s the establishment of a year-round station fulfilled the requirement but over the years the interpretation has shifted to accepting any substantial research on Antarctic issues, even without establishing stations or sending expeditions.

Some have argued that the desire of minimising environmental impacts argues for reducing human presence in Antarctica, including that of scientists. Thus, future research should be based not on sending more expeditions, but rather on cooperation using present infrastructure, and on investigations not involving Antarctic field studies. However, this argument does not take into account that, compared with other parts of the world, the Antarctic is still very much 'under-investigated', so that the main problem is not too many scientists, but rather not enough! Such an approach would also reduce the possibility of gaining new knowledge, e.g. from geologic structures not yet visited.

At the same time everybody agrees that duplicating infrastructure should be avoided, and that operations should constantly strive for a minimal environmental footprint. No other regions have such strict environmental standards as those now in operation in Antarctica. Over the years, a great variety of environmental issues have found resolution, such as establishing various types of protected areas,

Table 10.1 **Antarctic Treaty signatories**

Consultative Parties	Non-consultative Parties
Argentina	Austria
Australia	Belarus
Belgium	Canada
Brazil	Colombia
Bulgaria	Cuba
Chile	Czech Republic
China	Denmark
Ecuador	Estonia
Finland	Greece
France	Guatemala
Germany	Hungary
India	Korea (DPRK)
Italy	Malaysia
Japan	Monaco
Korea(ROK)	Pakistan
Netherlands	PapuaNew Guinea
New Zealand	Portugal
Norway	Romania
Peru	Slovak Republic
Poland	Switzerland
Russian Federation	Turkey
South Africa	Venezuela
Spain	
Sweden	
Ukraine	
United Kingdom	
United States	
Uruguay	

Notes: Original 12 signatories in bold.

regulating human behaviour (e.g. minimum flying heights above penguin colonies) and protecting various species. A multitude of fragmented regulations were eventually brought together in the Protocol on Environmental Protection that was signed in 1991. But before that was agreed there was high drama.

The dramatic 1980s

In the early 1970s the question of prospecting for mineral resources emerged, and came on the ATCM agenda in 1972. Following the oil crisis of 1973, some petroleum companies started seriously considering whether the continental shelves around Antarctica could contain commercial deposits of oil or gas.

SCAR and others were asked for advice, and the answer was generally that too little was known to estimate what could be there, but that all knowledge indicated that any commercial activity was unrealistic at present levels of technology, and would remain very unlikely for many decades.

After workshops and informal consultations for a few years, the ATCPs agreed in 1977 to a voluntary moratorium on mineral resource activities, which should last as long as progress was being made in negotiations of a regime. In 1982 formal negotiations started on a convention to regulate mining in Antarctica. At this stage the oil companies had lost interest, but it was still considered important to proceed in order to get regulations in place well before any commercial activity would be proposed.

To develop such an agreement seemed at the outset to be impossible, given that rights to the subsoil, which commercial operations required, meant finding a way forward which both claimant and non-claimant nations could accept. The approach used in article IV of the Treaty could not solve this issue.

A very different challenge to the negotiations came from a coalition of international organisations that launched a public pressure campaign against the process. They took the view that regulating such activities meant in principle accepting mineral activities, and they were completely against such developments in Antarctica. The negotiations went on for six years within what was termed the IVth Special Antarctic Treaty Consultative Meeting. This has so far been the only time when the delegates to an Antarctic Treaty meeting could experience walking through lines of demonstrating penguins, ranging in size from humans dressed in costumes to small inflatable ones. All carried signs protesting against the negotiations, and those who could, shouted slogans!

Despite these obstacles, the Convention on the Regulation of Antarctic Mineral Resource Activities (CRAMRA) was adopted in Wellington in 1988. In many ways this was a brilliant piece of political and legal diplomatic work in reconciling so many different viewpoints. It introduced stringent environmental standards, and required consensus before an area could be accessible for mineral activities. In that sense, it contained safeguards covering the needs of those who were opposed to such activities. Nevertheless the perception remained that mineral activities would now be easier, and so opposition continued.

The opposing views on sovereignty were resolved by an ingenious set of rules, which were sufficiently balanced that the parties could accept them. In this regard

Figure 10.5 Greenpeace demonstrations at the 3rd meeting of IVth SATCM, Bonn, July 1983, during the negotiations for the mineral regime. (Credit: Waltraud Demel, ASOC)

it was also important that the texts formalising their combined effects were not too explicit. It would not be acceptable, for example, to give special identified rights to the seven claimants in a way that would be in contradiction of Article IV. In this connection the most critical provisions were in the so-called regulatory committees, where the claimants could more easily establish a blocking situation than any other party.

The mineral discussions brought increased international attention onto Antarctica, with new interested nations taking what could be termed two separate lines of action. A number of countries decided they wanted to 'join the club', and started research activities in the Antarctic. These included the large European nations not already among the original 12, and most of the world's heavily populated countries. They recognised that reaching the position of an ATCP prior to the signature of the Convention meant in practice becoming permanent members of the Convention. This mirrors the way the original 12 contracting parties have permanent consultative status under the treaty.

A different approach was taken by a group of primarily developing nations led by Malaysia. In 1983 they introduced the 'Question of Antarctica' to the agenda of the United Nations General Assembly, claiming that Antarctica should be considered the 'common heritage of mankind' and therefore managed by the UN. The background for this initiative may have been a fear that arrangements would be reached that gave the developed nations the lion's share of any mineral resources. Similar fears had been

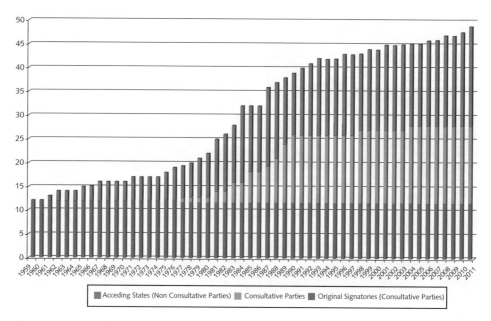

Figure 10.6 Growth in number of Antarctic Treaty Parties, showing the rapid increase in Consultative Parties in the 1980s associated with the negotiations on the minerals regime.

present with regards to the deep-sea bed, in the earlier negotiations on the Law of the Sea. Some of the participants from these also became involved in the mineral discussions, so there may have been a carry-over of such viewpoints.

The ATCPs decided they would not take part in any UN voting on Antarctica, as they considered that Antarctica did not need UN involvement. They maintained that the ATS and the ATCMs constituted a proven, well-functioning management of the region, to which any member of the UN could join. Over the years there has been no progress on this issue at the UN, and the interest there has gradually died down, in part because Malaysia who originally headed the campaign instead moved towards acceding to the Treaty.

The joy among the negotiators at having successfully completed the Convention did not last long. In 1989 Australia and France announced that they would not ratify it, rendering it, in reality, dead. There was probably a mixture of reasons for this decision. The one presented in public was a concern for the Antarctic environment, a view promoted by Greenpeace in a global campaign, and by the Cousteau Foundation in France.

For the first time in the history of the Treaty a situation had arisen where consensus had broken on a fundamental issue, and this presented a real possibility that the Treaty would fall apart. This was only 2 years before 1991, and as can be seen from article XII.2 any consultative party could after 30 years, i.e. in 1991, ask for a conference to discuss the Treaty, and if it was not satisfied with the results it could

Figure 10.7 New Zealand Ambassador Chris Beeby, who chaired the Antarctic Treaty Special Consultative Meetings on Antarctic minerals from 1982–88 and was a principal architect of the Convention on the Regulation of Antarctic Mineral Resource Activities. He died 19 March 2000. (Credit: Ministry of Foreign Affairs and Trade, New Zealand)

withdraw after a further 2 years. So tensions rose among the parties. Fortunately, all agreed that consensus had to be reached again, and no later than 1991.

The way forward involved negotiating a comprehensive regime to protect the Antarctic environment, including a mining ban. Given the short time available it was fortunate that much of the environmental regulations developed during the CRAMRA negotiations could be used.

The Protocol on Environmental Protection to the Antarctic Treaty was signed in Madrid on 4 October 1991 and entered into force in 1998 when it was ratified by the last of the ATCPs. It designated Antarctica as a 'natural reserve devoted to peace and science', and it set forth basic principles applicable to all human activities in Antarctica, including dispute settlement procedures. All activities relating to Antarctic mineral resources, except for scientific research, were prohibited for 50 years, i.e. until 2048. In this period the Protocol can only be modified by unanimous agreement of all Consultative Parties, and in addition, the prohibition on mineral resource activities cannot be removed unless a binding legal regime on Antarctic mineral resource activities is in force.

The Protocol has six Annexes: Environmental Impact Assessment (Annex I), Conservation of Antarctic Fauna and Flora (II), Waste Disposal and Waste

Figure 10.8 Antarctic Specially Protected Areas: photo of Cape Hallett with remains of Hallett Station in the foreground and the ASPA at the rear. (Credit: David Walton, BAS)

Management (III), Prevention of Marine Pollution (IV), and Area Protection and Management (V). The first four annexes were adopted in Madrid, together with the protocol. The fifth Annex was adopted only days later at the XVIth ATCM in Bonn, and entered into force in 2002.

Annex V replaced earlier categories of protected areas, and provides for the designation of Antarctic Specially Protected Areas (ASPA) and Antarctic Specially Managed Areas (ASMA). ASPAs are designated to protect outstanding environmental, scientific, historic, aesthetic or wilderness values, while an area where activities are being conducted or may be conducted in the future may be designated as an ASMA to assist in the planning and coordination of activities and minimise environmental impacts.

Annex VI on Liability Arising from Environmental Emergencies was the subject of protracted, complicated discussions. These issues of liability raised again fundamental legal questions, which had not been solved during the CRAMRA negotiations. Eventually the annex was agreed at the XXVIIIth ATCM in Stockholm in 2005. It deals with both governmental and nongovernmental activities in the Antarctic Treaty area. The operators of such activities must undertake reasonable preventative measures and establish contingency plans for responses to incidents which can adversely impact the Antarctic environment. They must take prompt and

Figure 10.9 At present the only Antarctic Specially Protected Species is the Ross seal. (Credit: Nigel Bonner, BAS)

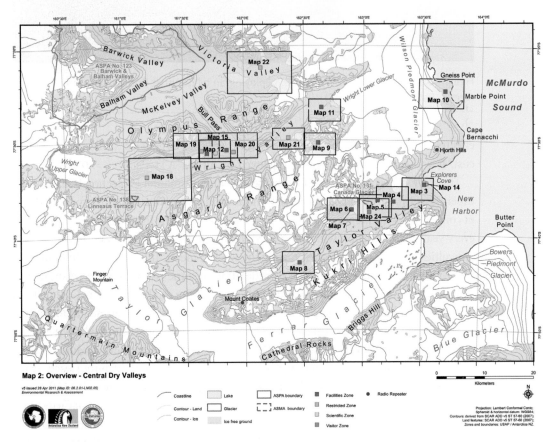

Figure 10.10 Antarctic Specially Managed Areas: map of the Dry Valleys. (Credit: AntarcticaNewZealand)

Figure 10.11 Marine pollution exercise, deploying an oil boom at Rothera Research Station. (Credit: Peter Bucktrout , BAS)

effective response action in case of environmental emergencies, or be liable for its cost. It will enter into force once approved by all the ATCPs.

The Protocol established the Committee for Environmental Protection (CEP) as an expert advisory body to provide advice and formulate recommendations to the ATCM in connection with the implementation of the Environment Protocol. The CEP meets every year in conjunction with the ATCM. Its workload has grown so fast that it has difficulties handling all the issues during the week it meets. It has therefore developed a multi-year work plan assigning full consideration of some issues to different years.

The adoption of the Environmental Protocol and the establishment of CEP in 1998 brought a new body of participants into the ATCMs, namely environmental managers. This has led to a clarification of roles and improved the interplay between SCAR and the ATCM. SCAR has always had, and still has, the role of giving scientific advice to the ATCMs, both through its own initiative, or by answering questions posed by the ATCM through SCAR's national representatives. This could be on any issue requiring knowledge of Antarctica, but increasingly over the years the questions dealt with the environment, especially related to protected areas. This led to SCAR establishing a Group of Specialists on Environmental Affairs and Conservation (GOSEAC) in 1988.

Ably filled with scientists of deep Antarctic knowledge, GOSEAC nevertheless frequently handled questions which were not so much science as management, sometimes causing friction within the Antarctic 'family'. Any blurring of roles in this respect was not only SCAR's fault. In the period before CEP, when there was also no Antarctic Treaty secretariat, it was very convenient for the ATCPs to ask SCAR with its competence and long collective memory to advise on practical environmental questions.

Looking now 20 years back at the high tensions of the 1980s, the question can be posed whether starting the minerals discussion was worth it, and whether once these were completed, was it for the better that consensus was broken?

The argument for starting the minerals discussion was one that frequently has been applied constructively in other Antarctic affairs, namely to treat an issue before it becomes an intractable problem. It would clearly have been even more difficult to negotiate a Convention on minerals after a site-specific commercial activity had been proposed. But as the view today is the same as two decades ago, namely that it will be many decades before any such activity is realistic, it may be argued that it was too early to spend so much effort at a non-existent challenge.

CRAMRA exists as a negotiated but not ratified convention. In this sense it may have considerable value should the issue surface in a more tangible way. And the negotiations on environment and liability helped paving the way for the Environmental Protocol and its annexes. So the negotiating efforts were by no means a complete waste.

Some will argue that the failure of CRAMRA was an illusory victory for the environment, for in reality it made the environmental protection weaker. This line of reasoning is that in a future changed political and industrial situation, the ATCPs could by consensus amend the mineral prohibition, and in that case there are few other constraints in place. Also under CRAMRA consensus was required to allow commercial mineral activities to proceed, and thereafter there were stringent regulations. Some of these have been replaced by the Environmental Protocol, but not all.

But these are hypothetical considerations, and it seems now unlikely that such a situation will arise. It can instead be said that the efforts during these years led to two other 'victories'. The first was that a much larger variety of nations came onto the Antarctic political scene, giving the ATCPs more global anchorage. Secondly, it taught the ATCPs that they could weather an internal crisis, and find their way back to consensus. That was an experience that should have long-lasting benefits.

Possibly this rediscovered cohesion helped in the solution of another difficult challenge, namely to establish a permanent secretariat for the Antarctic Treaty and the ATCMs. This was an issue that became more pressing during the 1980s, but with no resolution. During the following decade the critical question of location was not resolved; potentially it could be several places in the southern hemisphere. Argentina pushed strongly for Buenos Aires, and in 2001 they and the UK finally found an accord, which removed the latter's opposition. The Secretariat has been in full operation since September 2004. In addition to supporting the ATCMs, its roles include facilitating the exchange of information between the ATCPs and informing the public about the ATS. A full record of all considerations under the Antarctic Treaty is now readily available at http://www.ats.aq.

Box 10.1 Excerpts of the text of the Antarctic Treaty

Article I

1 Antarctica shall be used for peaceful purposes only. There shall be prohibited, inter alia, any measure of a military nature, such as the establishment of military bases and fortifications, the carrying out of military manoeuvres, as well as the testing of any type of weapon.

2 The present Treaty shall not prevent the use of military personnel or equipment for scientific research or for any other peaceful purpose.

Article II

Freedom of scientific investigation in Antarctica and cooperation toward that end, as applied during the International Geophysical Year, shall continue, subject to the provisions of the present Treaty.

Article IV

1 Nothing contained in the present Treaty shall be interpreted as:
   a a renunciation by any Contracting Party of previously asserted rights of or claims to territorial sovereignty in Antarctica;
   b a renunciation or diminution by any Contracting Party of any basis of claim to territorial sovereignty in Antarctica which it may have whether as a result of its activities or those of its nationals in Antarctica, or otherwise;
   c prejudicing the position of any Contracting Party as regards its recognition or nonrecognition of any other State's right of or claim or basis of claim to territorial sovereignty in Antarctica.

2 No acts or activities taking place while the present Treaty is in force shall constitute a basis for asserting, supporting or denying a claim to territorial sovereignty in Antarctica or create any rights of sovereignty in Antarctica. No new claim, or enlargement of an existing claim, to territorial sovereignty in Antarctica shall be asserted while the present Treaty is in force.

Article V

1 Any nuclear explosions in Antarctica and the disposal there of radioactive waste material shall be prohibited.

Article VII

1 In order to promote the objectives and ensure the observance of the provisions of the present Treaty, each Contracting Party … shall have the right to … carry out any inspection….

3 All areas of Antarctica, including all stations, installations and equipment …, and all ships and aircraft at points of discharging or embarking cargoes or personnel in Antarctica, shall be open at all times to inspection…

Box 10.1 **contd.**

Article IX

1 Representatives of the Contracting Parties … shall meet … at suitable intervals and places, for the purpose of exchanging information, consulting together … and recommending to their Governments, measures in furtherance of the principles and objectives of the Treaty …

2 Each Contracting Party which has become a party to the present Treaty by accession under Article XIII shall be entitled to appoint representatives to participate in the meetings … during such times as that Contracting Party demonstrates its interest in Antarctica by conducting substantial research activity there, such as the establishment of a scientific station or the despatch of a scientific expedition.

Article XII

1a The present Treaty may be modified or amended at any time by unanimous agreement of the Contracting Parties…

2a If after the expiration of thirty years from the date of entry into force of the present Treaty, any of the Contracting Parties … so requests …, a Conference of all the Contracting Parties shall be held as soon as practicable to review the operation of the Treaty.

b Any … amendment to the present Treaty which is approved at such a Conference by a majority of the Contracting Parties…, shall be communicated by the depositary Government to all Contracting Parties….

c If any such … amendment has not entered into force … within a period of two years … any Contracting Party may … after the expiration of that period give notice… of its withdrawal from the present Treaty; and such withdrawal shall take effect two years after the receipt of the notice by the depositary Government.

The secretariat is led by an Executive Secretary and operates under the direction of the ATCMs. The ATCPs are, so to speak, the Board of the Secretariat, but they only meet once a year, and there is no delegation of authority to a subgroup in the periods between ATCMs. In the early days it was clearly a challenge to find the right balance of roles between the Secretariat and the ATCMs. In particular the ATCPs do not want the Secretariat to be perceived as taking positions on contentious issues. The first Executive Secretary, Jan Huber from the Netherlands, had to operate within narrow bounds. To develop the trust of the ATCPs will no doubt be a target for its second chief officer, Dr Manfred Reinke from Germany, who took over in 2009.

The growth and management of fishing through CCAMLR

Article VI of the Antarctic Treaty states that nothing in the Treaty 'shall … affect the rights … of any State ... with regard to the high seas within that area'. In practice this means that with regards to the seas south of 60°S the ATCPs have dealt with questions by means of separate Conventions, with independent ratification and accession procedures.

There has been no significant seal hunting in the Antarctic since the nineteenth century. It was therefore easy for the ATCPs to develop the Convention for the Conservation of Antarctic Seals (CCAS), which was signed in London on 1 June 1972 and entered into force in 1978. So far, there has been little activity connected with this Convention.

The conclusion of CCAS, dealing with a High Seas resource, was part of the background for the consideration of a much broader and important issue, namely exploitation of krill and other marine resources. Both krill and trial fishing had started up in the Southern Ocean in the 1960s, krill harvesting grew rapidly, and there were indications that some stocks were overfished. In line with their precautionary approach, the ATCPs pushed for acquiring knowledge in order to regulate for sustainable exploitation. SCAR developed the BIOMASS programme in 1976, which together with other reports demonstrated the critical role of krill in the ecosystem. Clearly unchecked harvesting of krill could have grave impacts on whale, seal and bird populations.

The Convention on the Conservation of Antarctic Marine Living Resources (CCAMLR) was signed in Canberra on 20 May 1980 and entered into force in 1982. It provided for the conservation and rational use of krill, fin fish and other marine living resources in the Convention area. It did not cover whales and seals, for which there were other regulations in place already. There were 34 Parties to the Convention in 2011, including several who are not Parties to the Antarctic Treaty (Cook Islands, Mauritius, Namibia, Vanuatu and the European Union). All the original signatories to the Treaty are also members of the Convention.

The Convention area is not the same as that of the Antarctic Treaty, as the area is defined as being south of the Antarctic Convergence (Polar Front). This is a natural boundary between the cold waters of the Southern Ocean and the warmer waters to the north, and thus is also a boundary between different ecosystems. For long stretches the Convergence lies north of the 60th Parallel.

An innovative feature of CCAMLR is its ecosystem approach, which is based on evidence from marine biologists. In the evaluation of whether a species can be exploited the effect on its dependent species must also be considered. A great deal of scientific and technical work is done under the Convention, which is led by a commission with a scientific advisory body. It has its Secretariat in Hobart, Australia.

There have been tensions in CCAMLR both from internal and external forces. The first type include disagreements between those Parties that look on the Convention as a conservation measure, and therefore have wanted very small quotas, and the fishing nations who view the Convention, rightly, as one intended to set sustainable catch quotas.

The greatest challenges, however, have both been internal and external and relate to illegal, unregulated and unreported (IUU) fishing of Patagonian toothfish, *Dissostichus eleginoides*. This dense, white fish sells at high prices because of its quality. Harvesting of it became significant from the mid-1990s, and much of the catches have been taken by ships that have not adhered to the regulations.

In recent years the IUU fishing has come under better control through CCAMLR introducing a Catch Documentation Scheme (CDS), which makes it illegal for importers in their countries to buy this fish without CDS. As most of the high-paying consumers live in countries that are parties to CCAMLR, this has helped the situation, but there is still circumvention of the scheme, both by ships under a 'flag of convenience' i.e. a flag state not member of CCAMLR, and by ships flagged to CCAMLR Parties. The illegal fishing is generally done by companies registered in states that are not party to international fisheries agreements. In recent years many CCAMLR members have shown a strong willingness to enforce regulations, by cooperating in dramatic chases of ships conducting IUU fishing, with successful capture and prosecution, including confiscation of ship and catch.

CCAMLR is undoubtedly a success. The ecosystem approach with the precautionary principle means that the legal harvesting is well below sustainable levels, and is a model for other fishing conventions. Furthermore, various regulations have been put in place to address other problems, e.g. to prevent incidental bird mortality (being caught in lines and nets) and to reduce catching untargeted species.

Marine protected areas can be designated both under CCAS and CCAMLR. This is a challenge in relation to commercial activities. In such delineations care must be taken also when ATCMs designate specially protected areas with marine components, to ensure coherence within the Antarctic Treaty System.

Challenges ahead for the Antarctic Treaty and the management of the frozen continent

For a long time, tourism has been on the ATCM agenda, but there are few provisions that address only tourism and nongovernmental activities. The Parties differ in their views on tourism activity, and indeed in their direct management experiences with the industry. A minority of the Parties host tourism companies and/or provide port services. In a similar way at every ATCM there are opposing views between the industry, represented by their umbrella organisation IAATO (International Association of Antarctica Tour

Figure 10.12 Patagonian toothfish, the primary target of the illegal fishing in the Southern Ocean. (Credit: Paul Rodhouse, BAS)

Operators) and ASOC (Antarctic and Southern Ocean Coalition), the umbrella organisation of the environmental groups. Both have around 100 members, are based in the United States, and are invited to participate at the ATCMs. ASOC maintains that tourism should be restricted to reduce the environmental impacts, while IAATO maintains that tourism is a legal activity and that they are leading in their adherence to the Environmental Protocol. Some maintain that tourism is an activity for the pleasure of the individual, and so higher environmental standards should be set for such presence compared with science that advances public knowledge. Others maintain that tourism also functions as an important vehicle for positive promotion of Antarctica and its management, as those who go to the Antarctic generally are highly dedicated people.

Whatever the merit of these arguments, it can be seen that tourism increased greatly until 2007–08, when it reached a maximum with 32 000 visitors going ashore from cruise ships, and an additional 13 500 observing from overflying aircraft or cruise ships. The numbers have decreased strongly since then, presumably because of global finances. Recent prohibition of use of heavy fuels is also likely to reduce the number of

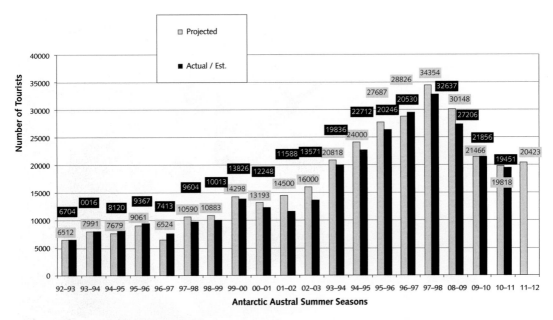

Figure 10.13 Tourist numbers over the last 20 years for those visiting the continent by sea. (Credit: IAATO)

large cruise vessels, which bring passengers without landing. In the 2010–11 season 19 400 went ashore, nearly all of these were ship-based tourists going to the Antarctic Peninsula, and 14 400 tourists went on board cruise vessels that did not make landings. There were no over-flights by aircraft from IAATO members. It was incidentally an over-flight that led to by far the largest accident in Antarctica so far, when a commercial airplane crashed into Mt Erebus in 1979, and all 257 on board were killed.

There have been several accidents with tourist vessels in recent years. This is perhaps not surprising, in view of the increased activity, and given that these waters are not all well charted, and frequently have icebergs and sometimes sea ice. The most dramatic accident so far was the sinking of *Explorer* in late November 2007. The 154 passengers and crew had to abandon ship, and lives could have been lost if cruise ships such as the MS *Nordnorge* had not been in the vicinity. There have also been cases of private yachts disappearing with loss of lives, and regulating small yacht activity is now another area of concern.

It should be noted that over the years there have also been several accidents both with ships and planes and helicopters operated by the national Antarctic research programmes, including two ships being sunk. So it is not just tourism that leads to such mishaps.

In recent years the ATCMs have included a special working group addressing the issue of tourism, but so far there has not been much progress on the most difficult issues. These include questions of establishing permanent tourist facilities on land, which once again raises the fundamental issues of rights to land. Who shall give a lease and a building permit?

Figure 10.14 MS *Explorer* sinking in Gerlache Strait in 2007. (Credit: Instituto Antartico Chileno)

Another challenge looming ahead is that of bioprospecting. Again it raises the issue of ownership. At present it is very unclear whether bioprospecting in Antarctica will be a significant commercial activity, but some maintain that the evolution of species in cold climates may have led to a gene pool of high significance. Exploitation at the microbial level is likely to cause little damage, but harvesting of target species, especially slow growing benthic ones, could be very damaging.

In summary it can be said that the problems in the near future for the Antarctic Treaty are likely to be new ones related to increasing commercial interests. Nevertheless, they seem smaller than those surmounted during the past decades. The ATCPs have shown an appreciation of scientific advice, and an ability to solve potential disputes before the problems became intractable. In recognising each other's fundamental limits the ATCPs have again and again been able to find accommodations that were acceptable by all sides within the safeguards provided for in Article IV. This ability to deal with issues before they become insoluble, coupled with a restraint in the pursuit of national interests, bodes well for the future management of Antarctica under a Treaty now over 50 years old.

11 | Antarctica: a global change perspective

ALAN RODGER

The natural world is the greatest source of excitement; the greatest source of visual beauty; the greatest source of intellectual interest. It is the greatest source of so much in life that makes life worth living. **David Attenborough**

Antarctica has been separated from the other continents for at least 30 million years. It is largely a frozen, barren continent with over 99% of its surface covered by ice, up to 4.5 km thick. It is the highest, driest, windiest and coldest continent and has the least vegetation and terrestrial life both in quantity and diversity. By complete contrast the Southern Ocean that surrounds Antarctica is rich in marine life, supporting many iconic species, including whales, seals, penguins and albatrosses. Antarctica has provided a remarkably stable environment over recent geological time and hence the physiological functions of many of its native marine species are now extremely sensitive to very modest changes in temperature. Antarctica is therefore considered to be one of the most fragile areas on Earth, and therefore most vulnerable to the effects of climate change.

Antarctica is at the very end of the Earth, yet human activity is having a significant impact on its environment. Changes in atmospheric circulation arising from the impacts of humans elsewhere are causing regional effects such as the break-up of ice shelves, marked changes in sea ice concentrations in the southern Pacific Ocean and Ross Sea areas, and reductions in krill concentrations in the South Atlantic. These are all associated with increased carbon dioxide emissions and the effects of the Antarctic ozone hole.

But what happens in Antarctica has global significance too. The most obvious examples are the melting of the ice, which contributes significantly to sea level rise, and changes in ocean circulation, responsible for moving 90% of the heat around the

Antarctica: Global Science from a Frozen Continent, ed. David W. H. Walton. Published by Cambridge University Press. © Cambridge University Press 2013.

planet driven by both the formation of sea ice and the interaction of the ocean with ice shelves. The Southern Ocean also has a large protein resource (krill) that will be exploited much further and faster as pressures increase on global food supplies, with consequent effects on the marine food web.

Current climate change in Antarctica

Both the northern and southern hemispheres have warmed significantly over the last century by about 0.6 and 0.8°C, respectively, with 2009 being the warmest year ever in the south. The difference between the hemispheres is mainly the result of the different distribution of land and sea, the latter having greater thermal capacity and thus warming more slowly. Globally 15 out of the 16 warmest years have occurred since 1995. There is considerable year-to-year variability, a significant element of which can be attributed to the effects of El Niño and La Niña. El Niño is a pattern of climate with approximately a 5-year periodicity whereby the eastern tropical Pacific Ocean temperature warms, and then cools (La Niña). This oscillation causes significant weather variations and extreme events, such as floods and droughts in many regions. In Antarctica, the major impacts are on sea ice distribution, which in turn impacts the breeding success of animals and birds as discussed later.

The warming is not uniform in either hemisphere with the Arctic and the Antarctic Peninsula showing rises well above average. The winter temperatures on the Antarctic Peninsula show a temperature rise of over 4°C in the last 50 years, a larger increase than in other seasons. Recent studies have demonstrated that these changes are outside the normal ranges of natural variability of the climate, and are consistent with the effects of human influences, partly owing to carbon dioxide increases and partly due the effects of the ozone hole. The rest of Antarctica shows some small variations but no statistically significant temperature trends. The caveat to this is that data sets are very short compared with those in other parts of the world and there are too little data to detect significant trends.

One very surprising recent finding is that the mid-troposphere (~5 km altitude) over Antarctica has been warming at a rate faster than anywhere else. The cause of the warming has not been determined. One suggestion is that it results from an increased occurrence of polar stratospheric clouds (~40 km altitude). These clouds form when the temperature falls below about $-78°C$ and their impact is to change the radiation balance through the atmosphere, trapping more incoming radiation in the lower atmosphere. Over Antarctica, the stratosphere is cooling, as the result of the combination of increased greenhouse gases and the ozone hole, making the formation of polar stratospheric clouds more likely.

Figure 11.1 Map shows the 10-year (2002–11) June, July and August (winter) average temperature change relative to the 1951–80 mean. The largest temperature increases are in the Arctic and the Antarctic Peninsula. (Credit: NASA Goddard Institute for Space Studies)

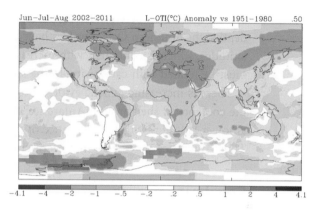

Changes in the Southern Ocean

The Southern Ocean comprises ~10% of the planet's oceans. It links the Pacific, Atlantic and Indian Ocean basins through the largest ocean current in the world, the Antarctic Circumpolar Current (ACC). The ACC (~125 Sverdrups) circulates around Antarctica and has a major role in isolating Antarctica from the warmer waters to the north.

More important in determining how the climate system of the Earth operates is the meridional over-turning circulation. This is where the cooling of surface waters and formation of sea ice around Antarctica result in an increase in the density of surface water, which sinks and then flows equatorwards carrying with it trace chemicals and nutrients along the sea floor. The Antarctic bottom water forming in the Weddell Sea flows northward beyond the equator in the Atlantic Basin. Warmer water returns to the Southern Ocean at intermediate depths. The same processes operate in the Arctic and together these convection-type circulations are responsible for the vast majority (~90%) of the poleward transport of heat on Earth and thus play a critical role in establishing the current global climate system. It is changes to these inter-hemispheric current systems that are claimed to be responsible for the see-saw action of the climate in the two polar regions. Ice core records show small-scale warming in Greenland occurred at the same time as a modest cooling in the Antarctic and vice versa. Changes in the seasonal production of sea ice are already occurring in both hemispheres but the records of ocean circulation are too short to show the expected changes, such as changing the strength of the currents at the bottom of the ocean.

The timescale of the meridional over-turning circulation is of the order of a thousand years but there is already evidence that some changes are occurring that are initiated in the surface waters. Over the last 80 years, the temperature of the Southern Ocean surface waters has risen on average by about 1°C, with those west of the Antarctic Peninsula and around South Georgia showing above average rises.

Warming has now been detected down to 3000 m below the surface. At intermediate depths, the Southern Ocean waters have also freshened, especially at the higher latitudes, which is consistent with increased ice melt, but such conclusions must be treated with care owing to the sparsity of data.

There is increasing genetic evidence to suggest that species developed on the cool, ocean shelf around Antarctica radiated equatorwards into the deep ocean at lower latitudes with the assistance of the meridional over-turning circulation. Thus Antarctica may be an important cradle for evolution and biodiversity.

Sea-level rise

Sea level at the last glacial maximum, about 20 000 years ago, was about 120 m lower than today. Water from the oceans was in the form of ice over the polar regions and higher middle latitudes. Over the intervening time, ice has melted and sea level has risen. About 100 years ago, the average global sea-level rise was about 1.8 mm per annum, but today the rate is four times greater. Melting of glacier ice at low and middle latitudes and thermal expansion of the ocean have been the most important factors, but now the rapid melting of the ice sheets in the polar regions is becoming much more significant. The Inter-Governmental Panel on Climate Change (IPCC) in 2007 identified the changes of cryosphere contribution to sea-level rise as the most significant uncertainty for future sea-level rise.

The Antarctic ice sheets contain over 10-times more ice than the rest of the Earth and hence its future is a critical component for sea-level projections. Observations of the changes in the ice mass have been made using radar, laser and now gravity measurement. These have been combined with data from ground and airborne surveys to provide a record of recent changes (~last 20 years) in the ice mass in the Antarctic.

The Antarctic ice sheet comprises three major components, the Antarctic Peninsula, West Antarctica and East Antarctica. The last is not changing markedly. Figure 11.2 shows there are some areas of very modest accumulation, which are counterbalanced with some areas of loss, especially at the exits to some glaciers, e.g. Totten and Cook glaciers in Wilkes Land.

The Antarctic Peninsula region has shown a sharp rise in temperature in the last half century. In summer, temperatures are now often above freezing and hence increased melting is occurring. The vast majority (87%) of the 244 marine glaciers are retreating, and the glaciers are flowing faster into the sea (a 12% increase in the decade from 1993). The Peninsula contributes about 5% of current global sea-level rise (0.16 ± 0.06 mm per annum). Some ice core records suggest that this acceleration may have started over 100 years ago but there is considerable uncertainly owing to the lack of instrumental records from the region.

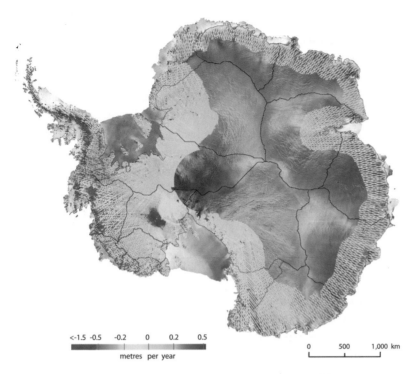

Figure 11.2 Recent changes in ice sheet elevation determined from ICESAT laser data (Pritchard, H. D., R. J. Arthern, D. G. Vaughan *et al.* (2009). Extensive dynamic thinning on the margins of the Greenland and Antarctic ice sheets. *Nature*, **461**, 971–975)

Over the same epoch, there have been reductions or complete loss of most of the major ice shelves on the Peninsula. Ice shelves do not directly contribute to sea-level rise as they are already floating, but they do have a buttressing effect on the glaciers feeding them. For example, since the Larsen B ice shelf on the eastern side of the Antarctic Peninsula broke up in 2002, there has been a marked (up to six times) acceleration of the glaciers that used to flow into the ice-shelf. Also there is now compelling evidence that the breakup of the Larson B ice shelf can be attributed to anthropogenic-induced changes to the pattern and strength of atmospheric circulation. The westerly winds around Antarctica have increased by about 15–20% in the last 30 years as a result of combined effects of increased greenhouse gas concentrations and the impacts of the ozone hole. This increased the warm air delivered to the eastern side of the Antarctic Peninsula and hence the melting and subsequent breakup.

The West Antarctic Ice Sheet (WAIS) is very different from the other two components of the ice sheet. There is an area at the southern end of the Antarctic Peninsula where there is a modest accumulation of snow. This is another impact of the changes in atmospheric circulation with more northerly winds bringing warmer, moist air to the region, increasing snowfall. But this modest increase is more than offset by the very rapid thinning (many metres per annum) of the glaciers feeding into the Amundsen Sea Embayment.

Since the initial thinning observations were made from radar altimetry data from space, there have been several coordinated international research activities to identify the causes of the changes. These have involved geophysical surveys from the air, on the ice and in the surrounding ocean to determine the present physical environment and the palaeohistory of the ice sheet. It has now been established that the recent changes of the WAIS result from changes in ocean currents. There is an increased flux of warm water being driven from the deep ocean up glaciated marine trenches in the continental shelf into the cavity beneath the marine glaciers of the WAIS. There is also compelling evidence that these ocean changes are a further result of changing atmospheric circulation patterns of anthropogenic origin.

Estimates of the current rate of mass loss from the Amundsen Sea Embayment vary by a factor of three but their average is about 100 Gt per year, equivalent to 0.3 mm per annum of sea-level rise or the current rate of mass loss from the entire Greenland ice sheet, and represents about 10% of current sea-level rise.

The studies have been extended across the whole of the continent. Over 20 of the 54 Antarctic glaciers studied, mostly those forming the WAIS, are now being melted from below by warm ocean currents.

One further reason for the major focus on the WAIS is that over most of the region the ice is resting on rock that is typically ~2000 m below sea level. Thinning of the ice near the marine edge causes the flow of ice to accelerate under the effects of gravity, leading to greater loss of ice and further thinning and hence contributes more to sea-level rise. Essentially, the ice sheet may become unstable, and the recent pattern of thinning could be a precursor to wholesale loss of this ice sheet, leading to an even more rapid sea-level rise ~1.5 m above current predictions. This is often referred to as a 'tipping point' – an abrupt, significant and irreversible change – and is a focus of current research effort. Since the IPCC 2007 report, the best estimate of sea-level rise by 2100 has been raised from 0.59 m to 1 m, but still with considerable uncertainty.

As well as the melting of ice shelves the pattern of melting is extending inland. Persistent melting – defined as melting that lasts for at least three daytime periods – has been steadily changing with not only the whole of the northern end of the Peninsula being affected but areas now in East Antarctica.

Many factors affect sea-level rise in addition to the melting cryosphere. About half the current rise results from thermal expansion as the oceans warm. Other factors include changes in the freshness of the water, a further consequence of

Figure 11.3 (a) Satellite data from 1987 to 2005 showing the first year in which persistent melting occurred from light green as 1987 to blue as close to 2005.
(b) Satellite data from 2005 showing how far inland persistent melting occurred with up to 50 days in the Peninsula and Marguerite Bay whilst even the Ross Ice Sheet and the area south of it was experiencing 10 days of melt. (Credit: NASA).

(a)

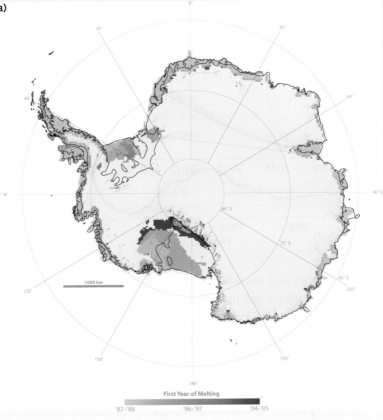

First Year of Melting

'87–'88   '96–'97   '04–'05

(b)

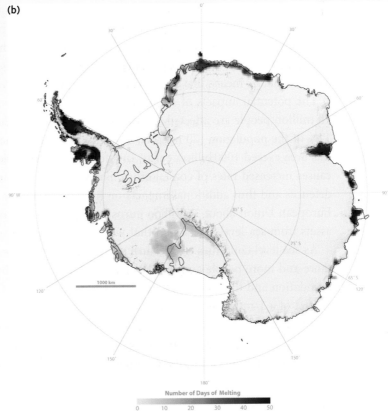

Number of Days of Melting

0   10   20   30   40   50

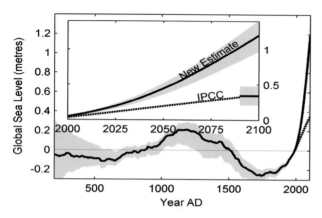

Figure 11.4 The rise in global mean sea level over the last 2000 years, with the projected rise to 2100 inset. The shaded regions are estimates of the uncertainty. (Grinsted *et al.*, 2009)

increased snow and ice melting, vertical motion of the land locally through tectonic processes and in higher latitudes isostatic recovery, sometimes called post-glacial rebound. During the last ice age (about 12 000 years ago) the polar and some mid-latitude regions were covered with thick ice. The weight of the ice caused the surface of the Earth's crust to sink. As glaciers have retreated since the end of the ice age, the load on the land has diminished and the depressed land is very slowly rising again. Greenland has lifted by about 70 m since the last ice age; rebound rates vary markedly and are not well determined across Antarctica.

Shorter timescale processes are super-posed on these longer ones. For example, changes in atmospheric pressure can change local sea level by ~1 m, albeit for a short time (days), seasonal run-off from the land can make changes of a further metre, and El Niño causes sea-level changes on a timescale of years. The combination of all these factors means that sea-level rise is not uniform across the world.

The potential impacts of sea-level rise are enormous. Every year about 10 million people are affected by coastal flooding. In the European Union, about 14% of the population (70 million) live within 500 m of the coast, and the assets in this area exceed 1000 billion euros. In addition to coastal flooding, rising sea level causes increased rates of coastal erosion, destruction of natural and artificial sea defences and thus additional impacts on humans and their livelihoods. In the European Union about 3.2 billion euros is spent annual protecting people and assets from sea-level rise, a sum that will continue to increase through the century.

As sea level continues to rise, and populations increase especially in urban areas, more and more people and infrastructure will be affected by flooding and inundation and thus the social impacts will be far greater. Current estimates of the impact of a 1-m rise in sea level will affect a few million to a few tens of million people in most continents. Asia is an exception; over 100 million people are likely to be affected, particularly in the low lying river deltas of the Ganges–Brahmaputra, Irrawaddy, Chao Phraya, Mekong, Yangtze and the Yellow River. Protection of

Figure 11.5 The lowest ozone values recorded at Halley Antarctic in October of each year. A Dobson unit is the thickness of the column of atmosphere if all ozone were compressed to 1 standard atmospheric pressure (~101 kPa) at 0°C, i.e. 200 DU = 2 mm.

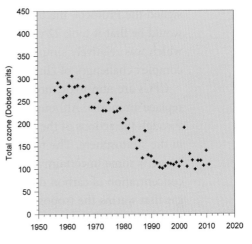

many low-lying areas will not be possible; managed retreat will be required, thus creating large numbers of environmental refugees. Accurate predictions of sea-level rise are essential to determine the optimum adaption strategies, and studies of the cryosphere in the polar regions have a critical contribution to make.

The ozone hole: lessons learned, problem solved?

The ozone hole was first described in an historic paper by British Antarctic Survey scientists in 1985 using measurements from Halley, Antarctica. The ozone hole is where most of the ozone in the stratosphere (~20 km altitude) over Antarctica disappears in the austral spring with a slow recovery in the summer. The minimum ozone concentrations observed in spring fell throughout the 1980s, and reached a minimum in the early 1990s but with considerable year-to-year variability. The finding of the ozone hole stimulated a considerable amount of new research, and its cause was quickly understood.

It is caused by chlorofluorocarbons (CFCs) used largely as a refrigerant and as a propellant in aerosols mainly in the northern hemisphere. But the largest impacts are seen in the most distant region of the planet. This is because the atmosphere below about 100 km altitude mixes very rapidly on the timescale of a year, thus polluting gases get transported quickly around the world. The Antarctic stratosphere, where the ozone hole forms, is much colder than anywhere else, making it an ideal location for the necessary chemical reactions to occur. The finding of the ozone hole was a sudden wake-up call to many people, as it demonstrated how humans could affect the planet in ways that were not predicted and through impacts that were not necessarily local.

What was equally remarkable was that within 28 months of the ozone hole paper being published, the Montreal Protocol on Ozone Depleting Substances was negotiated and signed by 24 countries, and the European Economic Community in September 1987. The Protocol was strengthened by a series of additional measures signed in London (1990), Copenhagen (1992) and Montreal (1997). This meant that CFC production was phased out in developed countries by 1995, and by 2010 for developing countries. The 16 September 2009 was an historic date – Timor-Leste signed the Protocol, the final UN country to do so. Another interpretation would be that it took 22 years for all countries to sign an international agreement which was relatively simple, and had little economic impact compared with the complex challenges of climate change today.

CFCs are non-toxic, non-flammable chemicals developed in the 1930s to replace ammonia. Although the concentrations of CFCs are just beginning to fall through the actions of the Montreal Protocol, they have lifetimes of ~20 to 100 years in the stratosphere. The recovery of the ozone hole is predicted to be around 2070. There is some uncertainty in the timing, as recovery is also affected by the concentration of carbon dioxide in the atmosphere. Carbon dioxide is a greenhouse gas that warms the troposphere, but acts to cool the stratosphere, and a colder stratosphere will delay the recovery of the ozone hole. The concentration of carbon dioxide through the twenty-first century is the largest unknown in climate change projections and hence the uncertainty.

The replacement for CFCs was initially hydrochlorofluorocarbons (HCFCs) but now hydrofluorocarbons (HFCs) are the major class of replacement substance. These chemicals have low ozone depleting potential but they are very effective indeed as greenhouse gases. The global warming potential (GWP) of carbon dioxide (100-year) is 1; the corresponding GWPs for the principal HFCs vary between 1000 and 4500. Specifically the GWP of HFC-125 is 3500, HFC-134a is 1430 and HFC-143a is 4470. Predictions suggest that the developing countries will increase their demand for HFCs by 800% by 2050. This is likely to add ~15% additional warming to that of carbon dioxide – equivalent to just a little less than the impact of the carbon dioxide produced today by the United States or China. There are no international regulations or agreements on HFC and HCFC gases despite the likely increase in demand for refrigeration in the future. There is a pressing need to find substances that have no ozone depleting potential and low global warming potential, as well as very low flammability and toxicity.

Impacts on the marine environment around Antarctica

The marine ecosystem of the Southern Ocean is often considered as relatively simple compared with those of other oceans. It covers not only 10% of the global seas but also a tremendous size range from microbial organisms (bacteria) to the largest animal that has ever lived on Earth, the blue whale.

A key element of the Antarctic marine ecosystem is the presence of sea ice, which provides higher predators with a platform for giving birth and for foraging, as well as a specialised habitat at the sea surface. Sea ice formation and its melting have a major influence on the timing, location, and intensity of biological production. Sea ice affects light and nutrient availability and ultimately food availability for

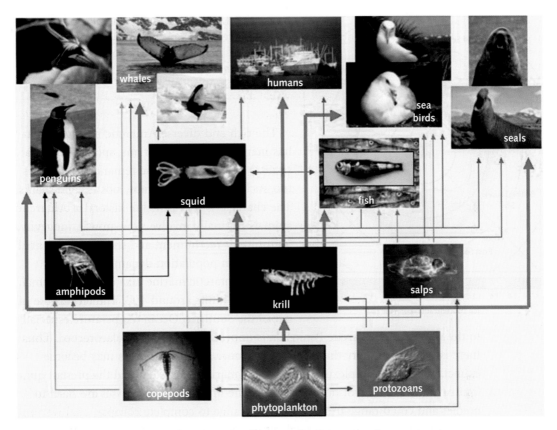

Figure 11.6 The Antarctic food web with krill being a central species in the system.

animals. The melting of sea ice in spring results in the stratification of the upper water column and release of nutrients that promotes primary production. Parts of Antarctica have extended periods of darkness and at times 24-hour daylight. As a result the Antarctic marine environment experiences many large seasonal variations, matched only in the Arctic.

The Southern Ocean marine environment changed fundamentally about 30 million years ago when the Drake Passage opened and the Southern Ocean cooled. Biota have since evolved in near-total isolation. They have had to survive the challenges of the Milankovitch cycles when, at glacial maximum, it is believed that many if not all the ice sheets extended to the continental shelf. There is compelling genetic evidence that many animals have persisted in Antarctic waters throughout this time and there is no geological evidence for climate-related extinctions of marine taxa on the timescales of the 23 000, 40 000 and 100 000 year Milankovitch cycles. Therefore, there must have been sufficiently large niches for the fauna to survive but how this was achieved remains a mystery.

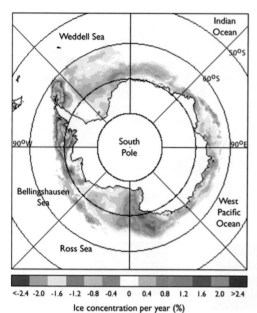

Figure 11.7 The variation in the concentration of sea ice in per cent per annum over the period 1979–2006. (Turner, J., R. A. Bindschadler, P. Convey *et al*. (2009). *Antarctic Climate Change And The Environment*. Cambridge: SCAR)

The rich and diverse Antarctic marine fauna has many taxa (e.g. cnidarians, sponges and sea squirts) but there are some remarkable absences too, such as brachyuran crabs, lobsters and sharks. The characteristics of the life history of Antarctic animals normally involve slow growth, longevity, intermittent recruitment and strong inter-annual variability in population dynamics.

Many Antarctic marine taxa are stenothermal, with upper limits around 4°C. The recent rate of rise of the Southern Ocean temperature especially to the west of the Antarctic Peninsula is unprecedented in the palaeorecord. Thus there is a real possibility that some of the more sensitive animals may become extinct. Given the complexities of even this marine ecosystem and the present quite limited understanding of how it operates, the impacts of such losses are hard to predict and could range from almost no change to complete collapse.

In addition to the strong natural cycles, humans have also had a significant role in shaping the Southern Ocean ecosystem, through the culling of the higher predators (whales, seals and penguins) over the last two centuries. The impact of humans could be described as causing a regime shift, whereby the system that was dominated by the great whales is now in a new state where the top predators are seals and birds.

The two main drivers of change in the marine environment in the twenty-first century are climate change and fishing. A central element of the Antarctic foodweb is krill (*Euphausia superba*). It feeds on phytoplankton. Some estimates suggest that there are about 600 000 billion krill in the Southern Ocean, and that they are the most abundant animal in the world. However, its life cycle is being affected by climate change through changes in sea ice concentrations.

The warming of the Antarctic Peninsula region is especially marked in winter. This, combined with the strengthening north-westerly winds, has resulted in a significant reduction (–6.8% per decade) in sea ice extent in the satellite era (since 1979). The change of wind patterns has also caused an increase in sea ice extent in the Ross Sea region (4.5% per decade), giving a small overall increase in sea ice extent around Antarctica (1% per decade). A significant proportion of the

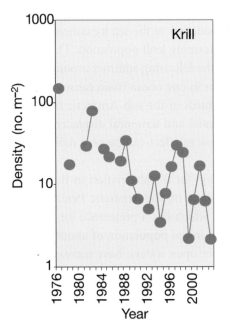

Figure 11.8 The change in the concentration of krill between 1976 and 2003. (Atkinson, A., V. Siegel, E. Pakhomov *et al.* (2004). Long-term decline in krill stock and increase in salps within the Southern Ocean. *Nature*, 432, 100–103)

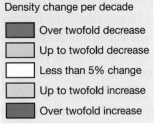

krill population overwinter in the Peninsula region feeding on the algae that grow on and in the sea ice. Also warmer winters favour smaller phytoplankton, such as cryptophytes; these are not consumed as effectively by krill. A larger population of cryptophytes means a smaller population of diatoms since the two species compete for nutrients. Smaller diatom populations mean less food for krill and hence less food for predators.

As the sea ice melts in summer the krill population advects into the South Atlantic where a very marked reduction in krill has been observed in the last 30 years. There has been a corresponding increase in the numbers of transparent tube-like creatures known as salps, animals which live in warmer, less food-rich areas. The impact of these changes has been reduced populations of the top predators around South Georgia.

There is considerable year-to-year variability in the concentration of krill in the vicinity of South Georgia. The variability appears to be a consequence of the El

Niño–La Niña climate oscillation. About 1 year following an El Niño event (warm water in the eastern Pacific), there is a reduction in the sea ice concentration around the Antarctic Peninsula and the over-wintering krill population. This impacts the food supply for the birds and seals the following summer around South Georgia. Thus a physical change near the equator in one ocean basin causes a major impact on the breeding success of animals and birds in the sub-Antarctic of a different ocean basin 2 years later. This large spatial and temporal displacement is a powerful illustration of how attribution of cause to effect can be very difficult in climate change studies.

Sea ice changes are also cited as the explanation for the alternation in the balance of the penguin population on Anvers Island on the Antarctic Peninsula. Since 1975, the numbers of the Adélie penguin, which has a preference for sea ice habitats, have fallen by a factor of eight from an initial population of about 16 000 birds. Gentoo and chinstrap penguins, that prefer open waters, have increased by a similar factor but from a much lower baseline figure. Hence there has been an overall reduction in pygoscelid penguins.

Sea ice is also critical in the life cycle of emperor penguins. It is now possible to identify their colonies from guano stains on the ice, and count individual birds using high resolution, multispectral imagery from space. These studies have found 10 previously unidentified colonies and one known colony has disappeared from the region where there have been large reductions in sea ice extent. The population of emperor penguins is calculated from the satellite data to be about 600 000 birds, about 60% higher than previous estimates.

So climate change, especially through the impacts on sea ice, is having a major impact on the top predators both through changes to the physical environment and to their food chain. Much less is known about changes of the lower trophic levels.

The terrestrial environment

As less than 1% of Antarctica is ice free, the range of terrestrial habitats is very limited indeed. Despite the small area, the logistic difficulties of reaching many exposed terrains means that Antarctica is poorly surveyed biologically and thus the patterns of dispersal are not well established. There is little species overlap between maritime and continental Antarctica in springtails, mites and nematodes and therefore unexpectedly high local endemism. Combined biogeographical, molecular, palaeological and geological studies are now all suggesting that much of the evolutionary separation events occurred tens of millions of years ago. This is an unexpected result. Glaciological modelling studies suggest that all the terrestrial environments, except for the Dry Valley region, were over-ridden at glacial maxima, and thus all terrestrial biology should have been very recent (i.e. since the last ice age), resulting from re-invasion from lower latitudes and hence little endemism. So the basic questions are: How did terrestrial ecosystems survive through the climate extremes of the glacial maxima? Where were the niches? How large do areas need to be for species to survive many tens of thousands of years, isolated from other locations?

There is a great deal yet to be done to explain how evolutionary and past processes formed present day polar biogeography and biodiversity. Knowledge of the life history of key species is also critical as this will allow determination of their capacity for adaptation at many levels – genomic, biochemical, cellular, physiological, structural and behavioural – in response to the changing biophysical environment.

Ecosystem services

Ecosystems provide goods and services that are fundamental to human wellbeing. Damage to the environment can seriously affect these 'ecosystem services' and can have significant economic implications too. The Millennium Ecosystem Assessment grouped ecosystem services into four dimensions:

- Provisioning services, such as food and bio-prospecting for pharmaceutical products;
- Supporting services, such as the draw-down of carbon dioxide from the atmosphere and the feedbacks on the climate system;
- Regulating services, such as the influence of and to the climate system, and
- Cultural services, such as education, recreation, tourism and aesthetic values.

The ambition of those involved in ecosystem services is to place a value on each element of the ecosystem. This grand challenge is, however, unlikely to achieve an international consensus on valuations. Whales are an obvious example; in parts of the world they would be considered from a conservation viewpoint, but elsewhere are a highly prized food source.

Over the last 200 years, the Antarctic has been providing food and other products. Exploitation of whales and seals was carried out extensively in the nineteenth and twentieth centuries resulting in these animals being driven to near extinction, while making a few companies and countries economically wealthy. Penguins were next on the list of species to be culled. In recent years, there has been further exploitation of the lower trophic levels of fish, and most recently krill which are being harvested for the nutraceutical and aquaculture industries.

The size of the krill resource is highly uncertain, estimates range from 133 to 500 million tonnes. The harvest in 2009–10 was 0.2 million tonnes, which represents a tiny fraction of the total krill biomass, so in principle the fishery is easily sustainable. However, the distribution of krill is highly variable with particularly high concentrations around South Georgia and on the coastal shelves, where fishing would be in direct competition with the seals and birds that breed nearby. This fishery, the least exploited in the world, is under mounting pressure year by year. At present, South Korea, Japan and Norway are the main nations involved in the krill fishery. New krill fishing initiatives have begun – Norway has recently commissioned two purpose-built trawlers and in 2010 China sent two ocean-going vessels to the Antarctic. During their 23-day exploratory voyage, they caught 2000 tonnes of krill, a daily catch of ~100 tonnes, three times greater than originally expected. This was a critical step in China's 5-year scientific research project called Development and Utilization of Antarctic Marine Living Resources, established by the Secretary for the Bureau of Fisheries.

Antarctic fisheries have already been severely impacted by humans. In the late 1960s, the Soviet Union targeted marbled notothenid and icefish around South Georgia. Over 400 000 tonnes were taken in the 1969–70 season followed by a rapid decline from which the stocks have still to recover fully.

In 1982, the Convention on the Conservation of Antarctic Marine Living Resources (CCAMLR) was established for all Antarctic marine living resources of the area south of the Antarctic Convergence. The novel element of CCAMLR was that 'living resources included the populations of fin fish, molluscs, crustaceans and all other species of living organisms, including birds' so it was the earliest fishery management organisation to take the entire ecosystem into account. The Convention also recognised the intimate relationship between the living resources and the physical environment long before climate change became a hot topic. The UN Food and Agriculture Organisation is now recommending that all fisheries follow the general principles of CCAMLR, i.e. by adopting a precautionary principle, and taking into account dependent species. Scientific advice and long-term monitoring data are essential to successfully implement an ecosystem approach to sustainable management of fisheries.

The focus of the Southern Ocean fishery in recent years has been the Patagonian toothfish which is a high value species. About 15 000 tonnes is caught annually with a retail value of around £300 million. This high value for fish has resulted in significant illegal and unregulated fishing. One of the serious weaknesses of the CCAMLR system is the inability to address illegal fishing – in one area it is suggested that 90% of the catch is by pirate fishing boats. This undermines attempts to conserve the fishery in the Southern Ocean.

Antarctica has considerable marine biodiversity, and probably much still to be discovered, especially in the deep sea. Much of it is endemic and has raised increasing interest for bioprospecting for novel organisms. The animals in Antarctica have evolved to withstand environmental extremes, such as cold temperatures and aridity, and may have enzymes that could be of considerable commercial value. The best-known work is on the antifreeze glycoproteins produced by various species of fish. Uses could include enhanced preservation of frozen foods and of tissues during medical operations. There is strong pressure to find cold-active enzymes for detergents that would work in cold water, thus saving considerable energy. At present, the Antarctic community and the Antarctic Treaty have not come to a consensus on how to address the many moral, commercial and legal issues surrounding bioprospecting. For example, uncertainty about the use and ownership of samples inhibits the engagement of industry in some countries in Antarctic research. However, the importance of bioprospecting from natural environments may be diminishing as more and more new products are designed in the laboratory.

Carbon dioxide has risen in the last 200 years to a level 30% higher than at any time in the last 800 000 years as determined from Antarctic ice cores. The rate of rise of carbon dioxide is 50-times faster now than at any time in the ice core record. The change in relative contribution of the isotope ^{13}C to ^{12}C over the last 200 years also confirms that the burning of fossil fuel is the main contributor, thus it is incontrovertible that the recent rapid rise in carbon dioxide is anthropogenic in origin.

Oceans absorb atmospheric carbon dioxide into the mixed layer, the thin layer of water at the top of the ocean with nearly uniform temperature, salinity and dissolved gases in two ways, physical absorption and via the 'biological pump'. Carbon dioxide slowly enters the deep ocean at the bottom of the mixed layer particularly in regions near the poles where cold, salty water sinks to the ocean depths. The oceans presently are responsible for removing 26% of the emissions of carbon dioxide. The ocean around Antarctica is the largest oceanic sink for carbon dioxide but it is now becoming less effective because the circum-Antarctic winds are increasing, by about 15–20% in the last 30 years. As a result there is increased turbulent mixing in the upper layers that brings more carbon dioxide rich water from depth to the surface, lowering the partial pressure between the ocean and the atmosphere hence reducing the flux of carbon dioxide passing into the water.

The biological pump is the process by which carbon dioxide is fixed by photosynthesis and then consumed by animals and ultimately some is transferred to the deep ocean in the form of dead organisms, skeletal and faecal material. The carbon occurs both in organic and inorganic forms where much of it is remineralised by the benthic community but some is sequestered on the ocean floor. In order to predict future carbon dioxide concentrations in the atmosphere, it is necessary to understand the way that the biological pump varies both geographically and temporally, as well as the impacts on the pump of changes in temperature, ocean circulation and ocean chemistry (e.g. acidification due to increased carbon dioxide). Much quantitative research remains to be done on this topic.

As carbon dioxide is dissolved in seawater, it forms carbonic acid. This process is called ocean acidification and is thought to have a major impact on the ability of many marine organisms to build their shells and skeletal structures. Since the beginning of the Industrial Revolution 250 years ago, seawater acidity has increased by 30%. If present trends continue, by 2060 the acidity of the oceans will be greater than anything experienced in the past 20 million years. The major research focus to date has been on the possible increased dissolution of skeletal material, as the current rate of change is many times faster than anything experienced in the last 55 million years and animals and plants may not be able to adapt to changing conditions. Major changes in phytoplankton assemblages are likely to occur too. Under increased carbon dioxide conditions, some species of phytoplankton will grow more effectively, whereas others will be less successful owing to the increased acidity. As phytoplankton underpin the entire ecosystem, major impacts are possible but the current level of understanding is low, so no reliable predictions are possible. Since much more carbon dioxide is dissolved in cold water compared with the warm tropical regions, the impacts in the polar regions of ocean acidification are likely to be much more immediate and more significant.

Increased acidification improves the propagation of sound – it could increase by as much as 70% by 2060. Thus whales and dolphins will find it harder to navigate and communicate as the ocean becomes noisier, natural noises will be louder and there is an ever-increasing number of anthropogenic sources including drilling, sonar and boat engines.

Antarctica is physically remote, inhospitable, unpredictable and potentially dangerous, but as such captures the imagination of young and old alike, through the heroic exploits of the early explorers, the amazingly beautiful landscapes, the near-pristine environment and the characteristics of iconic species, such as penguins, albatrosses, whales and seals. Antarctica is the world's largest wilderness and as a result has high aesthetic value in many cultures.

Tourism in Antarctica started in the 1950s, and the industry has grown steadily over the intervening decades with numbers reaching about 33 000 in 2010–11 arriving by sea, or air. Most of the tour operators are members of the International Association of Antarctica Tour Operators (IAATO), which has a strict, but voluntary, code of conduct to minimise the impact on the environment and the wildlife. Also currently most tourism is focussed on a very small number of sites.

Increasing travel of humans to Antarctica, both for science and tourism, combined with climate change, enhance the possibility of new species being introduced despite the climate being inhospitable. Perhaps the most challenging area concerns viruses, bacteria, yeasts, fungi and micro-algae. Globally these micro-organisms represent more than 90% of the species diversity, yet in Antarctica there are few measurements and no firm baselines from which to determine change, yet they are critical to ecosystem function and resilience. In the marine environment, the greatest risks are probably posed by the release of ballast water and the transfer of organisms attached to the hulls of ships. Now tour ships spend half a year in the Arctic and then 6 months in the Antarctic, increasing the risk of transferring species between the polar regions. Irreversible changes in the ecosystems of many sub-Antarctic islands have already occurred, e.g. through the introduction of rats and rabbits.

Albatrosses are regarded by many as iconic species and thus of cultural value, but their numbers are in serious decline. One of the impacts of long-line fishing is the incident mortality of birds (by-catch). Typically a long liner deploys ~10000 baited hooks during a single long-line haul. This attracts birds, such as albatrosses and petrels, and over the years thousands of birds have been caught and drowned. Conservationists and scientists have been working with the fishing industry to reduce the deaths. Measures include having streamers behind the fishing boats to prevent birds getting close to the hooks before they sink out of range of the birds' diving capabilities. These measures have meant that the by-catch of birds in the South Georgia area fell from ~6000 per annum in the late 1990s to none in 2006 and subsequent years. Alas, albatrosses are still on the decline; currently at the rate of 4% per annum for the wandering albatrosses. Research shows that birds are breeding just as successfully as previously, but the returns of birds to breed are falling. New tracking technology allows scientists to show that albatrosses often go to South American and South African waters to feed; in these locations the same by-catch mitigation measures have not been fully implemented.

Bi-polar comparisons and contrasts

The two polar regions have similarities. Both are much colder than the global average, largely covered by snow and ice, and contribute significantly to driving the large-scale ocean circulation. Both the Arctic and the Antarctic Peninsula have warmed at a rate about five times the global average over the last half-century. The icecaps on Greenland and the Antarctic Peninsula are both melting at lower altitudes as a result of there being many more days when the air temperature is above 0°C. Also sea ice trends in the Arctic are becoming more like that of Antarctica each year, i.e. dominated by sea ice that melts every summer with only a small area of multi-year ice.

There are many more differences than similarities between the poles and these contrasts can act as excellent tests for geophysical and biological models that have been developed in one hemisphere. Some of the differences include the geography. Antarctica is essentially land surrounded by ocean, whereas the Arctic is the converse but with substantial mountain chains at sub-Arctic latitudes. The

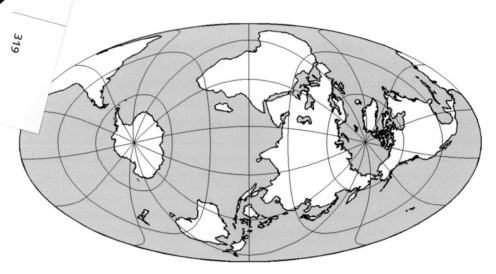

Figure 11.9 The two polar regions. The Antarctic is land surrounded by ocean whereas the Arctic is ocean surrounded by land.

consequences are many. There are indigenous peoples in the north but no permanent residents in Antarctica. There is land on which to deploy experiments in the south. The atmosphere in the north is much less stable and the troposphere and stratosphere in the north polar region have frequent incursions of warm mid-latitude air. Antarctica's climate system becomes quite isolated for extended times, especially in winter, and as a result the temperatures fall to the lowest on Earth, typically 20°C colder in the south than the north. One consequence is that ozone losses in the north are much smaller.

The Arctic, apart from two small areas of the high Arctic Ocean, is actively claimed by countries that surround it, thus there are many political constraints to carrying out research. The governance of the Antarctic is through the Antarctic Treaty System that has been in place for more than 50 years and enshrines Antarctica as being a continent set aside for science. Also, it is so large that no nation has the capacity to make all the necessary observations to understand how this component of the Earth system operates; this makes for ideal conditions to stimulate scientific collaboration.

Pollution from human activity and dust levels from the land are much higher in the Arctic than in the Antarctic, because the vast majority of the sources are in the northern hemisphere. So contrasts can be drawn between the pristine environment of Antarctica, and the Arctic atmosphere. As an example, the initial assessment and quantification of the role of the snow pack in the release of oxides of nitrogen (NO_x), and its impact on the concentration of OH radicals and ozone in the lower troposphere, were achieved in the south, where the atmospheric chemistry is simpler. Also black carbon (soot) levels are much higher in the Arctic compared with the south, being much closer to the sources; soot deposited on snow surfaces enhances the melting rate of snow and ice.

Even the Earth's magnetic field is very different in both polar regions. The magnetic pole in the north is close to the geographic pole and moving closer to it, by about 40 km per annum; in the south it is 24° displaced from the geographic pole and only moving about 10 km per annum away from the pole. These two very different configurations of the magnetic field have proved incredibly valuable in determining the way in which the Sun's particle energy enters the Earth's atmosphere and causes the aurora, and how these disturbances lead to space weather.

These major differences between the poles offer scientists both challenges and opportunities. For example, computer models developed in one hemisphere can be rigorously tested in the other hemisphere where the geophysical conditions are very different.

The future

The Antarctic has provided many critical insights into how the planet has operated in the past. For example, the ice core studies from many parts of the Antarctic covering nearly 1 million years have allowed the relationships between carbon dioxide and temperature variations to be established unambiguously.

Despite the best efforts of the many nations involved in Antarctic research, there is still a dearth of data and information. There is a pressing need for biodiversity mapping both on land and in the sea. There are large tracts that are not just under-sampled but unsampled. Sampling is vital to determine baselines now as some regions are already experiencing significant change. However, to ensure maximum value of any mapping programme, internationally agreed protocols are essential and organisations such as the Scientific Committee for Antarctic Research should play a leading role in developing these.

Making data that have already been collected available is equally important to allow studies of spatial and temporal patterns of physical, biological, glaciological and geological phenomena. Whilst there have been some notable successes, such as the Census for Antarctic Marine Life, much more needs to be done; a move away from providing catalogues alone and facilities to find and download data automatically is essential.

Environmental research in the twentieth century focused on understanding and quantifying individual processes, and often involved studying them in a small number of locations but now the focus is on looking at the entire Earth as a system, a complex series of interactions of many processes occurring on all timescales and all spatial scales.

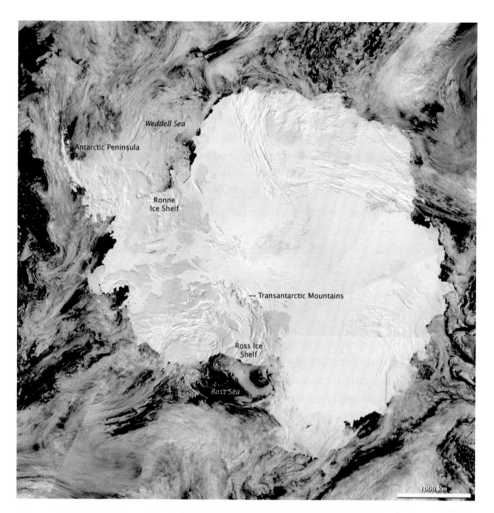

Figure 11.10 Antarctica: a major driver of global change in the atmosphere and in the oceans. (Credit: NASA)

Predictions of future climate change and its impacts require complex computer models that integrate more and more processes. There are still major processes occurring in Antarctica that are essential to understand and quantify if the present uncertainties in the model outputs are to be reduced. Perhaps the most important is the role of the Antarctic in the global ocean circulation and the transport of heat, nutrients and carbon amongst the ocean basins. Much of this is driven by sea ice formation and through interaction of the ocean with ice shelves. Whilst observations from satellites are excellent at producing global coverage of surface and atmospheric parameters, processes under the ocean surface are essentially invisible. Only very limited ocean data are available from *in-situ* measurements of moorings, gliders and from ships. These data show considerable variability of timescales of weeks, years and decades. Further developments are essential both to the models and data collection, with measurements in areas of ice shelves and in sea ice regions being particularly challenging.

The interface between rock and the continental ice is very varied both spatially and temporally – the ice can be frozen to the bed or be covered with till and water that makes the sliding resistance of ice orders of magnitude lower. Some remote sensing techniques have been developed, but validation of the technique is required through direct observations. This will require significant drilling of ice and with much greater spatial mapping of the ice–rock interface to allow better calculations of the rates of loss of ice from Antarctica and more accurate estimates of sea-level rise.

Presently the Earth experiences Milankovitch cycles of about 100 000 years which are related to the eccentricity of its orbit but prior to about 1.2 million years ago the period was 40 000 years, attributed to the obliquity of the Earth's orbit. The change has been observed in ocean sediments, but if old ice could be found in Antarctica it would provide carbon dioxide and other important chemical signals to help explain why the dominant period changed. Understanding the natural rhythms is critically important to allow the anthropogenic impacts to be accurately quantified.

Many of the outstanding issues are so large that they are beyond the scope of a single nation to provide, and hence international cooperation, already a very strong characteristic of those operating in Antarctica, will need to increase further. To address some of the issues associated with the Southern Ocean requires continuous circum-polar measurements of a wide range of physical and biological parameters. A plan has been developed entitled the Southern Ocean Observing System to address the problem but it requires considerable investment from many nations to achieve significant progress. As many parts of the world experience severe financial budget constraints, it will be even more important to coordinate and share the use of major assets in Antarctica, stations, ships and planes. These activities will require some relaxation of national agendas and a more cooperative approach.

The Antarctic Treaty System has proved to be very successful to date, but in the longer term it will come under increasing pressure as the natural resources on the other continents diminish. Although oil and gas are often discussed in an Antarctic context, the extraction of rare Earth minerals might be the first serious challenge to the Treaty although identification of their locations and their mining will prove extremely difficult. The mineral resource issues are decades away but they clearly need to begin to be addressed now before any crisis might arise.

The challenges facing humans today are often expressed as food, water and energy security and the relevance of Antarctica to these three issues is not obvious at first glance. Certainly there are marine food resources, such as krill, that could be harvested sustainably. But incorporation of Antarctica into Earth System models is essential to predict accurately future climate scenarios for the whole planet. Sea-level rise and ocean circulations are just two examples where the processes in Antarctica are fundamentally important globally. Accurate prediction allows humans to make the optimum climate adaptation and mitigation choices. Thus Antarctic is remote but more relevant to humanity than it has ever been.

Visiting Antarctica

For a long time Antarctica really was impossible to get to unless you were part of an official expedition or owned your own boat. The first tentative steps were taken in the late 1950s when Chile and Argentina took the first paying tourists onboard their naval vessels used in resupplying their research stations. The current form of expedition cruising with expert lecturers and landings to visit wildlife and heritage sites was pioneered by a Swedish–American Lars Eric Linblad whose first chartered ship cruised south in 1966. He went on to build the first ice-strengthened tourist expedition ship – the *Linblad Explorer* – and began regular cruises on this in 1969. Meanwhile adventurous yachtsmen had seen this as a new frontier and yachts from several countries began to visit the Antarctic and the sub-Antarctic islands at the end of the 1960s.

Airborne tourism began with flights from South America down over the Antarctic Peninsula in 1956 but it was not until 1977 that regular overflights began from New Zealand and Australia. Flights to the interior began in 1984 but it was not until the establishment of inland facilities with a connecting air bridge using a blue ice runway in 1987 that annual tourism of the interior of the continent began.

In 1991 the seven major tour operators decided to form a trade association and work together to promote responsible tourism. The International Association of Antarctic Tour Operators (IAATO) has since grown to over 100 members, encompassing almost all the principal tour operators into Antarctica. Antarctic Treaty Governments encourage tourists to use IAATO companies because of their interest in environmental protection, their careful adherence to safety standards and the extra dimension of both enjoyment and information they supply through their guest lecturer programmes. Their web site at http://iaato.org/ provides details of member companies and will direct you to the wide variety of tours and activities now available.

At present, around 40 ships provide the mobile bases for most Antarctic tourism. The majority of cruises are focussed on the Antarctic Peninsula, its offshore islands and the sub-Antarctic island of South Georgia. Most cruises leave from Ushuaia in Argentine Tierra del Fuego, and visit sites in the South Shetland Islands and down the west coast of the Peninsula. Some longer cruises go via the South Orkney Islands whilst some also visit South Georgia. There has been an

Antarctic circumpolar cruise normally every 2 years onboard an icebreaker but the future of this is linked to the banning of ships using heavy fuel oil in Antarctic waters. Occasional cruises go from South America via South Georgia and Tristan da Cuhna to Cape Town. A small number of cruises go each year to the Ross Sea, visiting the New Zealand sub-Antarctic islands en route, or to Commonwealth Bay to visit Mawson's hut, and leave from Bluff in New Zealand or Hobart in Australia.

Whilst some of the cruises are themed – either by language or by subject – most are not, with lectures primarily in English. For small parties who wish to visit some of the more inaccessible beaches or go climbing on the Peninsula or South Georgia there are small yachts for charter taking groups of up to 10 people. Some cruises now offer chances to SCUBA dive or camp ashore.

Some people want to see the interior and either climb the highest mountain (Vinson Massif at 4897 m high) or reach the South Pole. A long-established company Adventure Network International (www.adventure-network.com) and other companies such as The Antarctic Company (http://www.antarctic-company.info) provide air transport in to a blue ice runway, such as the one at Union Glacier, where they maintain a summer camp which supports climbing and skiing parties. Other commercial airlines organise Antarctic overflights either out of Melbourne and Adelaide, Australia or Ushuaia, Argentina.

All tourism is now undertaken under the regulations agreed for the Protocol to the Antarctic Treaty for Environmental Protection. This means that all tourist activities originating from countries that are Treaty members must undertake impact assessments and in many cases may require an annual licence to operate the cruises and tours.

How do you choose a cruise?

With a wide range of operators how can you decide which company and ship to go with? Firstly, make sure it is with an IAATO member company to guarantee a proper standard of environmental awareness and passenger information. Initially you need to decide where you would like to go – Peninsula, sub-Antarctic islands, Ross Sea – and if you would like a themed holiday say with a group of similarly minded people – bird watchers, for example. Then look to see the usual language used for lecturing – there are ships with English, Spanish, French, German and even Japanese lectures. Finally, look carefully at the price and what is included and excluded.

Some cruises visit historic huts or have permission to land you at a research station so that you can talk to Antarctic scientists and see something of the current research.

The timing of when you go may not be flexible but if you are keen to see young penguin chicks and seal pups then you need to go before Christmas. The weather is always very changeable but generally it is best in late December through to late February. The daily weather has a major influence on when and where you land so please read the small print and realise that the list of possible landings in the itinerary is just that – possible if the weather is right.

Equally important is deciding how big a ship to go on. The ships vary widely in age, capacity and levels of luxury so research the options carefully. Look at www.polarcruises.com to see details of the various types of ship available. The small ships of less than 200 passengers are much more intimate and you are likely to get to know many of the people on board as well as the lecturers. The larger ships (200–500) have greater facilities but, because of restrictions on passenger numbers on shore, they usually do fewer landings than the small ships. On the other hand, their larger size often makes them more comfortable in bad sea conditions and they almost always have more facilities! The largest cruise ships (above 500 passengers) simply cruise by without landing passengers anywhere.

Although there is no age limit on who can go, to be able to take advantage of the landings you will need to be reasonably agile to be able to get into and out of the landing boats. You can, of course, remain entirely on the ship throughout the cruise if this is a problem. Much has been written about the Drake Passage and its weather – it can be flat calm as well as being storm force 10 but it is normally only a 2-day crossing and sea sickness pills or patches are very effective for most people.

D. W. H. Walton

Further reading

The literature on Antarctica is considerable and there are many accounts of the early explorers and their expeditions. Much of the scientific material is published in research journals and is not easily accessible. Here we list books that the interested reader can reasonably expect to find on the internet or borrow through their local library. Only titles in English are listed but there are books in Spanish, French, German, Dutch, Russian, Chinese and Japanese on a variety of Antarctic subjects.

Chapter 1

Fogg, G. E. (1992). *A History of Antarctic Science.* Cambridge: Cambridge University Press, 508 pages.

Reader's Digest Association, Inc. (1985). *Antarctica: Great Stories from a Frozen Continent.* Sydney, Australia: Reader's Digest, 319 pages.

Riffenburgh, B. (2007). *Encyclopedia of the Antarctic*, 2 vols. New York: Routledge, 1272 pages.

Sale, R. (2002). *To the Ends of the Earth: The History of Polar Exploration.* London: Harper Collins, 224 pages.

Chapter 2

Anderson, J. B. (1999). *Antarctic Marine Geology.* Cambridge: Cambridge University Press, 289 pages.

Cooper, A. K., P. J. Barrett, H. Stagg, B. Storey, E. Stump, W. Wise and the 10th ISAES editorial team, eds. (2008). Antarctica: A keystone in a changing world. *Proceedings of the 10th International Symposium on Antarctic Earth Sciences.* Washington DC: The National Academic Press, 150 pages.

Faure, G. and T. M. Mensing (2010). *The Transantarctic Mountains: Rocks, Ice, Meteorites and Water.* Heidelberg: Springer Verlag, 804 pages.

Florindo, F. and M. Siegert (2009). *Antarctic Climate Evolution.* Amsterdam: Elsevier, 593 pages.

Futterer, D. K., D. Damaske, G. Kleinschmidt, H. Miller and F. Tessensohn (2006). *Antarctica: Contributions to Global Earth Sciences.* Berlin: Springer Verlag, 477 pages.

Gamble, J. A., D. N. B. Skinner and S. Henrys (2002). *Antarctica at the Close of the Millennium.* Wellington, New Zealand: Royal Society of New Zealand, 652 pages.

Stump, E. (1995). *The Ross Orogen of the Transantarctic Mountains.* Cambridge: Cambridge University Press, 284 pages.

Chapter 3

Alley, R., W. Broekner and G. Denton (2011). *The Fate of Greenland: Lessons From Abrupt Climate Change*. Cambridge, MA: MIT Press, 232 pages.

Bamber, J. L. and A. J. Payne, eds. (2004). *Mass Balance of the Cryosphere*. Cambridge: Cambridge University Press, 662 pages.

Benn, D. I. and D. J. A. Evans (2010). *Glaciers and Glaciation*. London: Hodder Education, 816 pages.

Cuffey, K. and W. S. B. Paterson (2010). *The Physics of Glaciers*, 4th edition. Oxford: Butterworth-Heinemann, 704 pages.

Bennett, M. M. and N. F. Glasser (2009). *Glacial Geology: Ice Sheets and Landforms,* 2nd edition. Oxford: Wiley-Blackwell, 400 pages.

Glushak, K. (2009). *The Surface Mass Balance in a Regional Climate Model of Antarctica.* Saarbrücken, Germany: VDM Verlag, 164 pages.

Hambrey, M. and J. Alean (2004). *Glaciers,* 2nd edition. Cambridge: Cambridge University Press, 394 pages.

King, J. C. and Turner, J. (1997). *Antarctic Meteorology and Climatology*. Cambridge: Cambridge University Press, 422 pages.

Marshall, S. J. (2011). *The Cryosphere*. Princeton, NJ: Princeton University Press, 312 pages.

National Research Council (2011) *Future Science Opportunities in Antarctica and the Southern Ocean*. Washington DC: National Academies Press, 230 pages.

Rudiman, W. F. (2007). *Earth's Climate, Past and Future*. London: W. H. Freeman, 480 pages.

Siegert, M. (2001). *Ice Sheets and Late Quaternary Environments*. Oxford: Wiley, 248 pages.

Solomon, S., D. Qin, M. Manning, *et al.*, eds. (2007). *Climate Change 2007: The Physical Science Basis. Contribution of Working Group I to the Fourth Assessment Report of the Intergovernmental Panel on Climate Change*. Cambridge: Cambridge University Press.

Thomas, D. N. and G. S. Dieckmann (2009). *Sea Ice*, 2nd edition. Oxford: Wiley-Blackwell, 640 pages.

Turner, J., R. A. Bindschadler, P. Convey *et al.*, eds. (2009). *Antarctic Climate Change and the Environment*. Cambridge: SCAR. Available for download at: http://www.scar.org/publications/occasionals/ACCE_25_Nov_2009.pdf.

Zotikov, I. A. (2006). *Antarctic Subglacial Lake Vostok: Glaciology, Biology and Planetology*. Berlin: Springer Verlag, 139 pages.

Chapter 4

King, J. C. and J. Turner (1997). *Antarctic Meteorology and Climatology*. Cambridge: Cambridge University Press, 422 pages.

Mawson, D. (1915). *Home of the Blizzard, Being the Story of the Australian Antarctic Expedition 1911–1914*. London: Heinemann, (2 vols) 438 pages.

Solomon, S., D. Qin, M. Manning, *et al.*, eds. (2007). *Climate Change 2007: The Physical Science Basis. Contribution of Working Group I to the Fourth Assessment Report of the Intergovernmental Panel on Climate Change*. Cambridge: Cambridge University Press.

Turner, J., R. A. Bindschadler, P. Convey *et al.*, eds. (2009). *Antarctic Climate Change and the Environment*. Cambridge: SCAR. Available for download at: http://www.scar.org/publications/occasionals/ACCE_25_Nov_2009.pdf.

Turner, J. and G. J. Marshall (2011). *Climate Change in the polar regions*. Cambridge: Cambridge University Press, 448 pages.

Chapter 5

Carmack, E. C. (1990). Large-scale physical oceanography of polar oceans. In W. O. Smith, Jr, (ed.) *Polar Oceanography*. San Diego, CA: Academic Press, pp. 171–222.

Hempel, G. and I. Hempel, eds. (2009). *Biological Studies in Polar Oceans: Exploration of Life in Icy Waters*. Bremerhaven: Wirtschaftsverlag NW, 352 pages.

Orsi, A. H. and T. Whitworth III (2004). *Hydrographic Atlas of the World Ocean Circulation Experiment (WOCE). Volume 1: Southern Ocean*. Southampton: International WOCE Project Office, available at http://woceatlas.tamu.edu/Sites/html/atlas/SOA_PRINTED.html.

Rintoul, S., M. Sparrow, M. Meredith, *et al.* (2012). *The Southern Ocean Observing System: Initial Science and Implementation Strategy*. Cambridge: SCAR, 74 pages.

Olbers, D., V. Gouretski, G. Seiß and J. Schröter (1993). *Hydrographic Atlas of the Southern Ocean*. Bremerhaven: Alfred Wegener Institute, 82 pages.

Siedler, G., J. Church and J. Gould, eds. (2001). *Ocean Circulation and Climate*. San Diego, CA: Academic Press, pp. 475–488.

Solomon, S., D. Qin, M. Manning, *et al.*, eds. (2007). *Climate Change 2007: The Physical Science Basis. Contribution of Working Group I to the Fourth Assessment Report of the Intergovernmental Panel on Climate Change*. Cambridge: Cambridge University Press.

Steele, J., S. Thorpe, K. Turekian, eds. (2001). *Encyclopedia of Ocean Sciences* (6 volumes). San Diego, CA: Academic Press, 3399 pages.

United Kingdom Hydrographic Office (2009). *Antarctic Pilot*, 7th edition. Taunton, UK: UK Hydrographic Office.

Webb, D. J., P. D. Killworth, A. C. Coward and S. R. Thompson (1991). *The FRAM Atlas of the Southern Ocean*. Swindon, UK: Natural Environmental Research Council, 67 pages.

Chapter 6

Agnew, D. (2004). *Fishing South: The History and Management of South Georgia Fisheries*. London: Penna Press, 128 pages.

Bergstrom, D. M, P. Convey and A. H. L. Huiskes, eds. (2006). *Trends in Antarctic Terrestrial and Limnetic Systems*. Berlin: Springer Verlag, 369 pages.

Denlinger, D. L. and R. E. Lee (2010). *Low Temperature Biology of Insects*. Cambridge: Cambridge University Press, 404 pages.

Knox, G. (2006). *Biology of the Southern Ocean*, 2nd edition. Baton Rouge, LA: CRC Press, 621 pages.

Kock, K.-H. (1992). *Antarctic Fish and Fisheries*. Cambridge: Cambridge University Press, 375 pages.

Rodgers, A. D., N. M. Johnston, E. Murphy and A. Clarke, eds. (2012). *Antarctica: An Extreme Environment in a Changing World*. Oxford: Wiley-Blackwell, 564 pages.

Sømme, L. (1995). *Invertebrates in Hot and Cold Arid Environments*. Berlin: Springer-Verlag, 275 pages.

Thomas, D. N., G. E. Fogg, P. Convey *et al.* (2008). *The Biology of the polar regions*. Oxford: Oxford University Press, 416 pages.

Vincent, W. F. and J. Laybourn-Parry (2008). *Polar Lakes and Rivers: Limnology of Arctic and Antarctic Aquatic Ecosystems*. Oxford: Oxford University Press, 320 pages.

Chapter 7

Clark, S. (2007). *The Sun Kings: The Unexpected Tragedy of Richard Carrington and the Tale of How Modern Astronomy Began*. Princeton, NJ: Princeton University Press, 224 pages.

Eather, R. H. (1980). *Majestic Lights: The Aurora in Science, History and the Arts*. Washington DC: American Geophysical Union, 323 pages.

Lanzerotti, L. J. and C. G. Park, eds. (1978). *Upper Atmosphere Research in Antarctica*. Antarctic Research Series, Vol. 29. Washington DC: American Geophysical Union, 264 pages.

Moldwin, M. (2008). *An Introduction to Space Weather*. Cambridge: Cambridge University Press, 146 pages.

Suess, S. T. and B. T. Tsurutani, eds. (1998). *From the Sun: Auroras, Magnetic Storms, Solar Flares, Cosmic Rays*. Washington DC: American Geophysical Union, 146 pages.

Chapter 8

Fowler, A. N. (2000). *COMNAP: The National Managers in Antarctica*. Baltimore: American Literary Press, 165 pages.

COMNAP Secretariat (2010). *Proceedings of the COMNAP Symposium 2010: Responding to Change through New Approaches*. Christchurch, New Zealand: COMNAP Secretariat, 120 pages.

Xu, S., ed. (2003). *Proceedings of the Tenth Symposium on Antarctic Logistics and Operations*. Beijing: China Ocean Press, 196 pages.

Chapter 9

El-Sayed, S. Z., ed. (1994). *Southern Ocean Ecology: The BIOMASS Perspective*. Cambridge: Cambridge University Press, 399 pages.

Fifield, R. (1987). *International Research in the Antarctic*. Oxford: ICSU Press and Oxford University Press, 146 pages.

ICSU/WMO (2007). *The Scope of Science for the International Polar Year 2007–2008* (WMO/TD-No.1364). Paris and Geneva: ICSU and WMO, 79 pages.

Krupnik, I., D. Hik, I. Allison, *et al.* (2011). *Understanding Earth's Polar Challenges: International Polar Year 2007–2008*. Alberta, Canada: CCI Press, University of Alberta, 950 pages.

SCAR (2010). Antarctic Science and Policy Advice in a Changing World: SCAR Strategic Plan 2011–2016. Special Report. SCAR: Cambridge, 34 pages.

Turner, J., R. A. Bindschadler, P. Convey, *et al.*, eds. (2009). *Antarctic Climate Change and the Environment*. Cambridge: SCAR. Available for download at: http://www.scar.org/publications/occasionals/ACCE_25_Nov_2009.pdf.

Walton, D. W. H. and P. D. Clarkson (2011). *Science in the Snow: Fifty Years of International Collaboration Through the Scientific Committee on Antarctic Research*. Cambridge: SCAR.

Chapter 10

Berkman, P. A., M. A. Lang, D. W. H. Walton and O. R. Young, eds. (2011). *Science Diplomacy: Antarctica, Science and the Governance of International Spaces*. Washington DC: Smithsonian Institution Scholarly Press, 337 pages.

Dodds, K. (1997). *Geopolitics in Antarctica*. Chichester, UK: John Wiley, 270 pages.

Griffiths, T. and M. G. Haward (2011). *Australia and the Antarctic Treaty System: 50 Years of Influence*. Kensington, Australia: University of New South Wales Press, 352 pages.

Joyner, C. C. (1998). *Governing the Frozen Commons: The Antarctic Regime and Environmental Protection*. Columbia, SC: University of South Carolina Press, 382 pages.

Liggett, D. H. (2009). *Tourism in the Antarctic: Modi Operandi and Regulatory Effectiveness*. Saarbrücken, Germany: Müller, 244 pages.

Støkke, O. S. and D. Vidas, eds. (1996). *Governing the Antarctic: The Effectiveness and Legitimacy of the Antarctic Treaty System*. Cambridge: Cambridge University Press, 486 pages.

Triggs, G. D. and A. Riddell, eds. (2007). *Antarctica: Legal and Environmental Challenges for the Future*. London: British Institute of International and Comparative Law, 437 pages.

Winther, J.-G., R. T. Andersen, B. L. Basberg and T. Bomann-Larsen (2008). *Norway in the Antarctic: From Conquest to Modern Science*. Oslo: Schibsted forlag, 239 pages.

Chapter 11

Cornell, S., I. C. Prentice, J. House and C. Downy (2012). *Understanding the Earth System: Global Change Science for Application*. Cambridge: Cambridge University Press, 328 pages.

Skinner, B. J. (2008). *The Blue Planet: An Introduction to Earth System Science*. 2nd edition. Chichester, UK: Wiley, 576 pages.

Visiting Antarctica

Carpentier, G. (2010). *Antarctica: First Journey. The Traveller's Resource Guide*. Ontario, Canada: Hidden Brook Press, 360 pages.

Naveen, R., C. Monteath, T. Du Roy and M. Jones (1990). *Wild Ice: Antarctic Journeys*. Washington DC: Smithsonian Books, 224 pages.

Rubin, J., ed. (2008). *Antarctica*, 4th edition. Footscray, Australia: Lonely Planet Guides.

Acknowledgements

David Walton is grateful for permission from Bill Manhire and June Back to reproduce the poems, for assistance from Ron Lewis Smith, Bob Burton and Pete Bucktrout with the photographs and for permission from John Kelly, Philip Hughes, Chris Drury and Alan Campbell to use their artwork. Victoria Wheatley and Kim Crosbie kindly provided comments on tourism.

Bryan Storey wishes to thank David Cantrill, John Smellie, David Vaughan, Andy Smith, Michael Studinger, Ian Dalziel, Antarctica New Zealand, Trond Torsvik, Robert Nicholls and the British Antarctic Survey for all their help with the illustrations for his chapter.

Funding from the United States National Science Foundation has supported John Cassano's Antarctic research. To my wife Liz and daughter Sabrina, thank you for your patience during my long absences while in Antarctica. None of this would have been possible without your love and support.

Louis Lanzerotti and Allan Weatherwax wish to acknowledge gratefully the National Science Foundation for supporting solar–terrestrial and astronomical research in Antarctica.

Lou Sanson is grateful for the assistance of Ursula Ryan, Joanna Eckersley, Peter Fretwell and Tom Riley in the preparation of his chapter.

Olav Orheim is grateful for the assistance of the Norsk Polarinstitutt, INACH, Raquel Beeby and David Gaston in obtaining the illustrations for his chapter. He wishes to thank Steve Wellmeier of IAATO, Jim Barnes of ASOC, Manfred Reinke and Tito Acero at the Antarctic Treaty Secretariat for useful comments on earlier versions of his manuscript. The chapter expresses his personal views only, and not institutions with which he has been associated.

Index

Page numbers in *italic* denote figures. Page numbers in **bold** denote tables.

Acaena magellanica 190
accessibility, through media 32–3
accretion, Antarctic Peninsula 59–60, *60*
acidification 318
Adélie Land 6, 276
aerosols, ice core data 84–6
air, density 115
air bubbles, ice matrix *95*, *96*
air transport
    research stations 238–40, *239*, *241*
    weather forecasting 128–9
aircraft 240
Alaskozetes antarcticus 197
albatross 170–1
    by-catch 170–1, *171*, 179, 319
    creative art *28*, *29*
    wandering *170*, *171*
albedo *104*, 104–5
    ocean 150–1, *151*
    sea ice 69–74, *74*, 150–1, *151*
algae 163–4, 166, *190*
    see also macroalgae
alien species introduction 204–5, *208*, *209*, 318–19
amphipoda 183, *186*, 186
Amundsen Abyssal Plain 139
Amundsen, Roald 11
    race to South Pole 12
Amundsen Sea Embayment, glacier thinning 306
Amundsen-Scott South Pole Station 230–1, *232*, *233*
    hydroponics 236, *237*
ANDEEP project 184–5
ANDEEP-SYSTCO project 185–6
ANDRILL 37
angiosperms *see* plants, flowering
annexation 13–18, 275–7
Antarctic Biodiversity Database 265
Antarctic Bottom Water 122, *145*, 147, 153, 163
Antarctic Circumpolar Current 80, 137, 139–40, 148–9, *149*, 303
    variability 155
Antarctic Circumpolar Wave 155
Antarctic Coastal Current 139–40
Antarctic Data Directory System 263

Antarctic Digital Database (ADD) 265
Antarctic Digital Magnetic Anomaly Project 265
Antarctic Intermediate Water 146
Antarctic Map Catalogue 265
Antarctic Master Directory 263
Antarctic Peninsula 38
    barrier winds 116–20
    fossils 50–1, 62–5
    ice sheets and climate change 304, *307*, 312–14
    maritime terrestrial ecosystem 193
    plate tectonics 42
    subduction and accretion 59–60, *60*
Antarctic Plate 42, *43*
Antarctic and Southern Ocean Coalition (ASOC) 297–8
Antarctic Specially Managed Areas (ASMAs) 290, *291*
Antarctic Specially Protected Areas (ASPAs) 290, **290**
Antarctic Specially Protected Species *291*
Antarctic Treaty 17, 22–3, 280–2, 294–5
    Article IX 284
    Committee for Environmental Protection 292
    conservation of flora and fauna 283
    Consultative Meetings (ATCMs) *281*, 282–4
    Consultative Parties (ATCPs) 282–4, 287, **288**
        management of Antarctica 288
    Convention on the Conservation of Antarctic Marine Living Resources (CCAMLR) 296–7
    Convention for the Conservation of Antarctic Seals (CCAS) 296
    Convention on the Regulation of Antarctic Mineral Resource Activities (CRAMRA) 286–7
    environmental management 292–3
    evolution 282–5
    Measures 283
    mineral resource exploitation 286–7, *287*, *289*

    Protocol on Environmental Protection 232, 284–5, 289–92
    SCAR 266–7, *267*, **268**, 282–3
    Secretariat 293–5
    tourism 297–9, *299*
Antarctic Treaty System 204, 283, 323
Antarctica
    common heritage of mankind 287–8
    comparison with Arctic 319–21, *320*
        temperature 93, 112–13, 161–2
    the future 321–4, *322*
    global importance of 210
Apatosaurus 50
Apple huts 244, *245*
Araucaria 65
Arctic
    comparison with Antarctica 319–21, *320*
        temperature 93, 112–13, 161–2
Argentina
    claims over Antarctica 15–17, 276–7
    Laurie Island station 10, 276
art 27
    creative programmes 28–32
Artedidraco skottsbergi 10
Artists and Writers Program, National Science Foundation 28
Astrolabe Glacier *76*
astronomy 222–6
atmosphere
    composition, ice cores 86, *88*
        see also air bubbles
    global circulation 106–9
        climate change 301
        three cell model *107*, 107–9
    humidity 120–4
    pressure systems *139*, 139
        and sea-level rise 308
    temperature 109–10, *110*
        and density 115
    measurement 220
aurorae 216
Aurora Australis 211, *214*
    measurement 220
Australian Antarctic Expedition 1911–1914 12
Australian Antarctic Territory 14, 276

Australian National Antarctic
    Research Expeditions 18
Australian-Antarctic Basin 139
Automatic Geophysical Observatories
    219, 220
Automatic Weather Stations *118*,
    118–19, *119*

BANZARE 14
Basler DC-3 aircraft 240, *241*
Beardmore Glacier, flora 65
BEDMAP 265
Beeby, Christopher *289*
Belgica antarctica 197
Belgica expedition *9, 9*
Bellingshausen, Fabian Gottlieb
    Thaddeus von, exploration
    4–5, 273
Bellingshausen Plain 139
benthic ecosystems 179–89
    continental shelf 179–84
    deep sea 184–6
    evolutionary radiation 182–3
Bentley Subglacial Trough 41
beryllium, ice core dating 86
bi-polar comparisons 319–21, *320*
    temperature 93, 112–13, 161–2
biodiversity 304, 312, 316–17
    deep-sea benthos 184–6
    global importance of 210
BIOMASS Programme 24, 260–1, *262*
bioprospecting 300, 316–17
birds 167–72, 194–5
black smokers 59
blizzards 116
Borchgrevink, Carsten, *Southern
    Cross* expedition 9–10
Bottom Water, formation 153
Bouvet Island, maritime terrestrial
    ecosystem 193
brachiopods *180*
Brachiosaurus 50
Bransfield, Edward, exploration 1820
    273
Bransfield Strait 42
Britain
    claims over Antarctica 14–17,
        276–7
    conflict with Argentina and Chile
        16–17
British Antarctic Expedition
    1910–1913 11
British Antarctic Survey
    Halley VI module 231–2, *234*
    ships 239

British Imperial Trans-Antarctic
    Expedition 1914–1916 12
British National Antarctic Expedition
    1901–1904 10, *13*
Bruce, Dr William, Scottish National
    Antarctic Expedition 10
by-catch 170–1, *171*, 179, 319
Byrd, Richard 17
Byrd Station, VLF radio
    measurement *221*
Byrd Subglacial Basin 41

Cape Adare, *Southern Cross*
    expedition 10
Cape Denison 116
Cape Hallett *290*
carbon cycle 79–80, 91–2
carbon dioxide
    acidification 318
    atmospheric 80, *134*
    biological pump 151, 317
    role in climate change 310
    sinks 75, 151, 317
    Southern Ocean 151
        impact on biogeochemical cycle
        154
carbon sinks 75, 79–80
    see also carbon dioxide, sinks
carbon sources 80
Carboniferous, Gondwana 47–8, *48*
Carpenter, William, oceanographic
    expedition 8
Casey Research Station,
    decompression chamber 248
Catch Documentation Scheme,
    CCAMLR 297
Charcot, Jean, French Antarctic
    Exploration 11
Cherenkov radiation 225
Chile, claims over Antarctica 15–17,
    276–7
Chinese Small Telescope Array
    (CSTAR) 226
chlorine, reaction with ozone 130–2
chlorofluorocarbons (CFCs) 130–2,
    309–10
circulation *see* atmosphere, global
    circulation, ocean circulation
Circumpolar Deep Water 146–7
climate
    and atmospheric CO_2 80
    evolution 88
    global, role of Southern Ocean
        150–1
    global models 133–4

greenhouse–icehouse transition 61
    near-surface temperature *111*,
        111–15, *113*
climate change 92–101, *97, 98*,
    128–35, 205–8, 302, *322*
    bi-polar comparison 319
    future prediction 136, 322–3
    ice cores 81–96
    impact on ecosystems 208–10
    impact on fauna and flora 182
    ozone hole 129–32
climate forcing 86
clothing, field camps 246, 246–7,
    247
coal measures, Permian,
    Transantarctic Mountains 48,
    49
Coleridge, Samuel Taylor, *Rime of the
    Ancient Mariner* 25
collaboration
    early 253–4
    future 323
Colobanthus quitensis 196, *198*
colonisation, terrestrial ecosystems
    202–4, *203*
communications, field stations 249
compasses, magnetic 2
Composite Gazetteer of Antarctica 265
Concordia Station *96*, 120, 232, *234*
conifers 50, *64*
continental shelf
    benthic ecosystem 179–84
    pelagic ecosystem 166
continental terrestrial ecosystem 193
Continuous Plankton Recorder 265,
    266
convection, ocean water 144
Convention on the Conservation of
    Antarctic Marine Living
    Resourse (CCAMLR) 296–7,
    316
Convention for the Conservation of
    Antarctic Seals (CCAS) 296
Convention on the Regulation of
    Antarctic Mineral Resource
    Activities (CRAMRA) 286–7,
    293
Cook, Captain James *3*
    Royal Society expeditions 3
coral *180*
coronal mass ejection 215
Cortada, Xavier, creative art *32, 32*
cosmic microwave background
    (CMB) 226
cosmic ray detection *223*, 223–5

Council of Managers of National Antarctic Programmes (COMNAP) 257
crabs 183–4
creative arts programmes 28–32
Cretaceous, flora 63
Cryolophosaurus ellioti 50
CRYOSAT, mapping sea ice thickness 73
cryoturbation 194, 195
cryptoendolithic communities 196, 200
CTD probe 156–8, 158
Cucumaria 180
currents 75, 75, 146–50
  climate change 306
  effect of ice sheets 93
  effect of sea ice 163
  role in energy budget 105–6
cyanobacteria 196, 198, 200

Dalrymple, Alexander, proposed southern continent 3
Dansgaard-Oeschger events 92–3
Dante robot 46
dating, stable isotopes 84, 86
Davis, Captain John, first landing 4, 274
de Leiris, Lucia, creative art 31
Debenham, Frank, The Quiet Land 26
decapoda 183–4
Deception Island 5
  Argentine claim 15–17
  volcanism 42–3, 44
  whaling 274
decompression chambers 248, 248
Degree Angular Scale Interferometer (DASI) 226
Deschampsia antarctica 196, 198
deuterium 81–4
Dicroidium 50
dicynodonts 50
dinosaurs 50–1
Dissostichus eleginoides, illegal fishing 297, 298, 316
diving 248, 248
dog sleds 37, 243, 243
Drake Passage 137
  opening 56–7, 80, 139
    impact on biota 311
Drake Plate 42
drilling 36, 37, 38, 81–96, 85
  oldest ice core 86–7

Dronning Maud Land
  first international scientific expedition 18, 277–80, 278
  Norwegian claim 14, 276
Dronning Maud Land Air Network (DROMLAN) 238–9
Dry Valleys see Victoria Land, Dry Valleys
Drygalski, Erich von 10
Dumont d'Urville, Captain Jules Sébastien César, French expedition 5–6
dykes 51, 56
  Marie Byrd Land 57
  Saunders Island 36

Earth
  energy budget 102–6
  orbital variation 85, 88, 89, 103–5, 323
East Antarctica
  Gamburtsev Mountains 40–1
  ice sheet 77–8, 80
  SWEAT hypothesis 46, 47
East Base 17
East Wind Drift 139–41
eccentricity 90, 323
ecosystems
  global importance of 210
  services 315–19
    cultural 318–19
    provisioning 315–17
    regulating 318
    supporting 317
  see also benthic ecosystems, pelagic ecosystems, terrestrial ecosystems
Eights, James
  Glyptonotus antarcticus 4
  scientific voyages 4
El Niño events 302
  impact on krill 313–14
  impact on sea ice 73
  sea-level change 308
El Niño-Southern Oscillation (ENSO) 156
electrical currents, ionosphere, mapping 220
Ellsworth Mountains 38, 39
  microplate rotation 57
Eltanin Fracture Zone 139
Enderby Abyssal Plain 139
Endurance, Ernest Shackleton 12, 14
environmental management, Antarctic Treaty 292–3
enzymes, cold-active 316–17

EPICA Dome C, ice cores 87–9, 88, 90
Epimeria angelikae 186
Euphausia crystallorophias 166
Euphausia superba 5, 312–14
eurybathy 188
evaporation, stable isotopes 84
evolution, marine taxa 304, 311
evolutionary radiation 182–3
expeditions 34
  first international scientific 18, 277–80, 278
  Royal Society of London 2–3
exploration
  early 1–8, 273–4
  Heroic Age 8–13
  imperialism 13–18
Explorer, sinking 299, 300
extinction, end-Permian 48–50

Falkland Island Dependencies 13, 276
  microplate rotation 57
  Survey (FIDS) 17
fata morgana 114, 114
Ferrel cell 107, 109
field camps 242–7
  accommodation 244, 245
  clothing 246, 246–7, 247
  food 246
  helicopter transport 240, 242
  waste 242
field stations, communications 249
Filchner, Wilhelm 11
fish
  evolutionary radiation 183
  fossil, Transantarctic Mountains 47
fisheries 178–9
  CCAMLR 296–7, 316
  icefish 316
  krill 168, 178–9, 315–16
  long-line, danger to wildlife 170–1, 171, 179, 319
flooding, effect of sea-level rise 308–9
food
  field camps 246
  research stations 236–7
Foraminifera 188
fossils
  Dicroidium 50
  fish, Transantarctic Mountains 47
  Glossopteris 48, 49
  invertebrate 51
  plant 62–5
  reptile 50–1
  species diversity 51
  Victoria Land Dry Valleys 64, 65

frazil ice 68
French Antarctic Expeditions 5–6, 11
freshwaters 200–2
    ecosystems 200–1
frontal systems
    water masses 147–8, *148*
        Southern Ocean 148
Frontoseriolis 186
Fuchs, Sir Vivian *20*, 20
fuel, research stations 237, *238*
fumaroles, Mount Rittmann 61

Gamburtsev Mountains 40–1, *41*
    see also Lake Vostok
Geodetic Control Database 265
geology 35–66
geomagnetic mapping 218–20
geophysical surveys 35–7, *38*
Gerlache, Lieutenant Adrien de,
        Belgica expedition 9
German Antarctic Expedition
        1901–1903 10
German Antarctic Expedition
        1911–1912 11
glaciation
    Carboniferous 47–8, *48*
    impact on terrestrial biota 192
    initiation 80
glaciers, thinning 305–6
global circulation *see* atmosphere,
        global circulation
global climate modeling 133–4
global warming 132–5, *133*, 302
Glossopteris 48, *49*
glycoproteins, antifreeze 316–17
Glyptonotus antarcticus 4
Gondwana 35
    birth 46–7
    breakup 56–7
        microplate displacement 57–8
    Carboniferous 47–8, *48*
    flora and fauna 47–51
    Mesozoic 50
    palaeogeography 47, *54*
    Permian 48–50
    plate tectonics 57–60
    subduction 59–60, *60*
    volcanism 51–7, *57*
Goodchild, H., Scottish National
        Antarctic Expedition *12*
grasses *191*, 196
gravity surveys 35–7
greenhouse gases, and climate
        change 86, *88*, 91–2, 132–5,
        134, 310

greenhouse–icehouse transition 61
    flora 62–5, *63*
Greenland, ice core data 92–3
grounding lines 79
    movement 98–9
    West Antarctic, Last Glacial
        Maximum 94
Group of Specialists on Environmental
        Affairs and Conservation
        (GOSEAC) 292–3
Grytviken *14*
    introduced plant species *209*
    Whaling Station *275*
guano *172*, 172
gyres, subpolar 149–50

Haag Nunataks 57, *58*
Hadley cell 107, 109
Hagglund tractors 239, *242*
Hallett Station *290*
Halley, Edmond, voyages to Southern
        Ocean 2
Halley Research Station
    drifting snow 117, 231
    ozone hole 129
    VI module 231–2, *234*
    waste *235*
helicopters 240
helioseismology *224*, 224
Hillary, Sir Edmund 20
HMS *Challenger*, oceanographic
        expedition 8
HMS *Chanticleer*, South Shetland
        expedition 5
HMS *Lightening*, oceanographic
        expedition 8
HMS *Porcupine*, oceanographic
        expedition 8
hollow earth theory 25
Hooker, Sir Joseph Dalton 7
    Flora Antarctica 7
hot spots
    biological 176
    mantle plume *53*, 56
Hughes doctrine 14–15
Hughes, Philip, creative art *30*, 31
human impact 204–5, 210, 312
    biological introductions 204–5,
        208, *209*, 318–19
    fisheries and bioprospecting
        315–17
    pollution 320–1
humidity 120–4
huskies *19*, *37*, *243*, 243
hydrochlorofluorocarbons (HCFCs) 310

hydrofluorocarbons (HFCs) 310
hydroponics, research stations 236, *237*

IBCSO bathymetric chart 265
ice
    crystals *95*
    global water cycle 67–80, *78*
        ice core data 84
    marine 145
    see also frazil ice; ice sheets;
            pancake ice; sea ice
ice cores 38, 81–96, *85*
    deepest 86–7
ice flow 94–8, 100
ice sheets *68*, *69*, 80
    age of ice 80
    air bubbles *95*, 96
    Carboniferous 47–8, *48*
    climate change 304–6, *305*, *307*
    dynamics 78–9
    East Antarctic 77–8, 80
    effect on ocean circulation 93
    formation 78–80
    mass balance 97–9
    subglacial water *99*, 99–100
    surface height *97*, 97–9
    thickness and volume 77, 94, *97*,
        97–9
    West Antarctic 77–8, 80
ice shelves 77–8
    climate change 305
    oceanic flow beneath 144–5, *146*
    thinning 97–8, *98*
ice streams 78–9, 100
ice–rock interface 323
icebergs *79*, 79, 145
icebreakers, research station supply
        238
IceCube neutrino detector *225*, 226
icefish *166*, 166, 316
imperialism *see* annexation
Indian–Antarctic Ridge 139
Inexpressible Island 122
insects, sub-Antarctic 195–6
instrumentation 156–9, *157*, *158*
International Arctic Science
        Committee (IASC) 257
International Association of
        Antarctica Tour Operators
        (IAATO) 297–8, 318, 325
International Council of Scientific
        Unions (ICSU) 254
International Geophysical Year
        1957–58 19–23, *21*, *23*, 253–4,
        280

International Geophysical Year
 1957–58 (cont.)
  solar–terrestrial research 218
  stations 20–2, *22*
  upper atmosphere studies 213
International Polar Years *24*, 24, 254
  Fourth 2007–2009, SCAR 267–70,
  270
inversion 109, *113*, 113–15, *114*
invertebrates
  fossil 51
  marine, dispersal 186–8
  terrestrial ecosystem 195–6
ionized gas 215–17, *217*
ionosphere
  electrical current mapping 220
  radio communications 213, 216,
  217
  SuperDARN radar 220
IPCC *see* United Nations
  Intergovernmental Panel on
  Climate Change
iron fertilization 154
islands
  biological hotspots 176, *177*, *178*
  ecosystems 192–3
Isoetes 64
isopoda 183, *186*, 186
isostasy, post glacial rebound 97, 308
isotopes
  as climate indicator 81–4, *85*
  and dating 84
  and evaporation 84

Jamesway huts *230*, 230
Japanese Antarctic Expedition
  1911–1912 11
jellyfish *165*
Jubany Station 232

Kelly, John, creative art *31*, 31
kelp 188–9, *189*
kenyte 43
King George Island
  Geographical Information System
  (KGIS) 265
  human impact *208*
krill 164, *165*, 167–8, *311*, 312–14
  baleen whales 174–5
  BIOMASS programme 24, 296
  Charles Wilkes 5
  commercial fisheries 168, 178–9,
  315–16
  and El Niño–La Niña oscillation
  313–14

ice 166
  impact of sea ice variability
  312–14, *313*

La Niña 302
Lake Fryxell *201*
Lake Vostok 41, *42*, 100, 201
lakes 200–1, *201*
  ecosystems 200–1
  subglacial 41, *42*, 99, 100, 201–2
  SCAR Programme *261*
land transport 239–43, *242*, *243*
Larsen B ice shelf, break-up 98, 305
Larsen, Captain C.A., Norwegian
  sealing and whaling 13, 274
Last Glacial Maximum 94
Laurie Island station 10, 276
Laws, Richard, creative art 31
Leptonychotes weddelli 4
Liability Arising from Environmental
  Emergencies 290–2
lichen 196, *198*, 200
Liothyrella uva 180
Little America III 17
longwave radiation 103, 105, 113–15
Louis-Philippe Land 6
Lystrosaurus 50

McMurdo Station *37*, *207*, 229–30,
  230
  decompression chamber 248
  food 237
McMurdo volcanic province 43, 61
Macquarie Ridge 139
macroalgae 188–9, *189*
Madrid Protocol 66
magma lake, Mount Erebus 46
magnetic field 2
  bi-polar comparison 321
  interaction with solar wind *215*,
  215
  studies 211–12
magnetic surveys 35–7
magnetosphere 215–16, *216*
Maitri Station, food 237
Manhire, Bill, *Erebus Voices – The
  Mountain* 27
mantle plumes 53, 56
maps, early 1, *2*
Marie Byrd Land
  volcanism *57*, 60–1, *61*
  West Antarctic Rift System 60–1
Marine Biodiversity Information
  Network (MarBIN) 265
marine ecosystems 162–89

food-web *311*, 312–14
  impact of climate change 310–14
Mario Zucchelli Station
  decompression chamber 248
  food 237
maritime terrestrial ecosystem 193
Markham, Sir Clements 8
Maudheim expedition 1949–52 18,
  277–80, *278*
Maury, Matthew, early oceanography
  7–8
Mawson, Sir Douglas 12, 14
  Australian Antarctic Expedition
  1912–1913 116
  The Home of the Blizzard (1915)
  103
Mawson research station 230
Maxwell Davies, Peter, *Antarctic
  Symphony* 32
medical care 129, 247–9
megaherbs *191*, 196
Mesozoic, Gondwana 50
meterology, Dr William Bruce 10
methane 93
  atmospheric variation 91
microbes 196, *198*, 200
microplates, rotation and migration
  57–8
mid-ocean ridges 138–9
midges 195, *197*
Milankovitch cycles 311, 323
mineral resources 66, 286–7, *287*, 323
mites 195–6, *197*
modeling
  global climate 133–4
  numerical weather prediction
  125–8
Montagu Island, volcanism 59
Montreal Protocol on Ozone
  Depleting Substances 131–2,
  309–10
mosasaurs 50
mosses 196, *198*
Mount Berlin 61
Mount Bird 61
Mount Discovery 61
Mount Erebus
  Dante robot 46
  volcanism 43–6, *45*, 61, *62*
Mount Hampton *61*
Mount Melbourne 61
Mount Morning 61
Mount Rittmann, fumaroles 61
Mount Sidley 61
Mount Terror 61

mountains, geological record 35, *36*, 37–41
Murray, John 8
music 32
Myosaurus 50

nacreous clouds *see* polar stratospheric clouds
National Science Foundation, Artists and Writers Program 28
navigation, early 2
nematodes 195–6, *197*
Neumayer, Dr Georg von 8
Neumayer III Station 231, *234*, *271*
neutrinos, AMANDA experiment *225*, 225–6
neutron monitoring *223*, 223–5
New Zealand, claims on Ross Dependency 276
Nordenskjöld, Otto, Swedish Antarctic Expedition 10
North Weddell Ridge 138–9
Norway, claims on Antarctica 14, 276
Norwegian Antarctic Expedition 11
Norwegian Whaling Expedition 13
Norwegian-British-Swedish expedition 1949–52 18, 277–80, *278*
Nothofagus 63–5
Notothenia larseni 10
notothenioids *165*, *166*, 166, 183, 316
numerical weather prediction models 125–8
nunataks 37
    see also Haag Nunataks

obliquity 90, 323
observation
    astronomical 222–6
    Southern Ocean 156–9
observation platforms 158–9
ocean circulation 146–50
    and biodiversity 304
    and climate change 303–4, 306
    effect of ice sheets 93
    effect of sea ice *75*, 75
    meridional over-turning 303–4
    role in energy budget 105–6, 303–4
ocean water
    convection 144
    density 141, 144, *145*
    temperature and salinity 141, *142*, *142*, *143*
oceanography, early 7–8
octopus *180*

oil exploration 66
open ocean ecosystem 176–9
Operation Highjump 17, 277
Operation Tabarin *16*, 16–17
orca 174, 176
Orcadas *see* Laurie Island Station
Ortelius, Abraham, World Map *2*
ozone
    absorption of UV solar radiation 109–10, *110*
    reaction with chlorine 130–2
ozone hole 129–32, *131*, *132*, *309*, 309–10

Pacific-Antarctic Ridge 139
Pagothenia borchgrevinki 183
Palmer, Nathaniel, exploration 1820 273
pancake ice 68, *71*
Paradiochloa flabellata 190
Peale, Titian Ramsay 5
peat 194
Pegasus white ice runway 238, *239*
pelagic ecosystem 162–79
penguins 167
    Adélie 167–8
        sea ice variability 314
    chinstrap, sea ice variability 314
    emperor 167, *169*, 169
        sea ice variability 314
    gentoo *178*
        sea ice variability 314
    sea ice variability 314
Permanent Service for Mean Sea-level (PSMSL) 265
Permian
    extinction event 48–50
    Gondwana 48–50, *49*
personnel
    selection 249–50
    training *250*, 250–2
    women *251*, 251
Peter I Island
    discovery 4
    maritime terrestrial ecosystem 193
petrels 167–8, 176
    by-catch 179
    snow *12*, 167–8, 172
Phaeocystis 166
photography 27
phytoplankton 166, 176, *177*
    upwelling 147–8
Pine Island Glacier, sub-glacial volcano 46

plants
    flowering 50, 62–5, 196, *198*
    fossil 48–50
    greenhouse–icehouse transition 62–5, *63*, *64*
    terrestrial ecosystems 196
plasma 215–17, *217*
plasmapause 217, 221
plasmasphere 216–17
plate tectonics 35, 42–3, 57–60
    impact on fauna and flora 182
    see also Gondwana
PLATeau Observatory (PLATO) 226
plesiosaurs 50
Pleuragramma antarcticum 166
Poa cookii 191
Poe, Edgar Allan, *Narrative of Arthur Gordon Pym of Nantucket* 25
poetry 25–6
Pointe Géologie 6
polar cell 107, 109
polar stratospheric clouds 130, *131*, 302
polar vortex 130
poles, magnetic 2
    see also South Magnetic Pole
pollution 290–2, *292*
    bi-polar comparison 320–1
polychaete *165*
polynyas 69, 122, *123*, *152*, 152–3
    formation of ice 75, 122, 152–3
Polytrichum 198
Port Martin station 18
Prasiola crispa 190
precession 90
precipitation 120–2
    distribution *121*
    formation 109
predators 176
primary productivity 147–8, 164
Princess Elisabeth Research Station 235–6, *236*
Pringlea antiscorbutica 191
prolacertids 50
pterosaurs 50

Rac-tents 244, *245*
radar, SuperDARN 220
radiation *see* evolutionary radiation; longwave radiation; solar radiation
radiation belts 217, *218*
radio waves
    communication 213, *217*
    VLF, measurement 220–1, *221*

radiosondes *124*, 124–5, *125*
rebound, isostatic 97, 308
REference Antarctic Data for Environmental Research (READER) 265
reindeer *208*
remote sensing 35–7, *38*
reptiles, fossil 50–1
research stations 16–18, 229–52
    design and construction 230–2
    drifting snow 117, 231
    environmental issues 232
    food and drink 236–7
    fuel 237, *238*
    hydroponics 236, *237*
    IGY 20–2, *22*
    impact on environment 204, *207*, *208*
    insulation 230, *231*
    international 232
    recreation 237
    renewable energy 235–6, *236*
    transport 237–43
        air 238–40, *239*
        land 239–43, *242*, *243*
        sea 239
    waste *235*, 235
research vessels 156–8, *157*, 239
rivers 200
Robin, Gordon de Q. *278*, 280
Robinson, Kim Stanley, *Antarctica* 25
Rodinia 46, *47*
Ronne Antarctic Research Expedition 17
Ronne-Filchner Ice Shelf 77–8
Ross, Sir James Clark 3, 6–7, *7*
    magnetic field studies *212*, 212
Ross Abyssal Plain 139
Ross Dependency 14, 276
Ross Ice Shelf 77–8
    discovery 7
    Southern Cross expedition 10
Ross Island *15*
    Mount Erebus 43–6, *45*
    volcanism 61
Ross Sea
    Sir James Clark Ross 3
    Southern Cross expedition 9–10
Rothera Research Station
    decompression chamber *248*, 248
    gravel runway 240, *241*
Royal Society of London, expeditions 2–3
runways
    blue ice 240, *241*

gravel 240, *241*
    white ice 238, *239*
Rutford Ice Stream *39*

Sabine, Edward, magnetic field studies 212
Sabrina Automatic Weather Station *118*
safety 247–9
salinity, ocean water 141, *142*, *143*
sastrugis *83*
satellites
    oceanographic data 159
    weather data 125, *126*, *127*
Saunders Island, volcanic dyke *36*
sauropods 50
SCAR (Scientific Committee on Antarctic Research) 24, 254–9
    achievements 260–1
    Action Groups and Expert Groups 256–7, **258**, 292–3
    Antarctic Treaty 266–7, *267*, **268**, 282–3
    BIOMASS Programme 24, 260–1, *262*
    conferences 257–9, *259*
    design of buildings 230–1
    Feature Catalogue 265
    Fellowship Programme 257
    first meeting 254, *255*
    Fourth International Polar Year 2007–2009 267–70, *270*
    GOSEAC 292–3
    National Members 255–6, **256**
    Presidents 257
    Prince of Asturias Prize 261, *262*
    products 263–5
    Research Programes 256, **258**
    science and policy 264–7
Schytt, Valter 277, *278*
science, collaboration, early 253–4
science fiction 25
science programmes, international 23–4
Scotia Ridge 5
Scott, Captain Robert Falcon 10–11
    race to South Pole 12
    tents 244, *245*
Scott Base *231*
    food 237
Scottish National Antarctic Expedition 10, *11*, *12*
sea anemone *180*
sea cucumber *165*, *180*
sea ice *71*, *76*, 151–3

albedo 69–74, *74*
    collapse *73*
    ecosystem 163–4
    extent *73*
    formation 67–75, *70*, *71*
        and ocean circulation *75*
    importance to marine ecosystems *75*, 162–4, 310–11
    thickness, CRYOSAT mapping *73*
    variability *73*, *312*
        impact on krill 312–14, *313*
        impact on penguins 314
sea-level rise 100, 304–9, *308*
sea urchin *180*, 185, *187*
sea-spider *180*
sea-squirt *180*
sealers 178
    early exploration 3–4
sealing 273, *274*
seals 172–4
    Convention for the Conservation of Antarctic Seals 296
    crabeater 6, 172–4, *173*
    elephant 159, 172, 178, 273, *274*
    fur 172, *173*, 178, 273
    leopard *176*, 176
    oceanographic data transmission 159
    Ross 7, 172, *291*
    Weddell 4, 172
sedimentary processes 98–9
sedimentary rock, West Antarctic Rift System 61
Seismic Data Library System (SDLS) 265
seismic surveys 35–7
seismicity 42–6
Shackleton, Sir Ernest 11–12, 14
    Cape Royds hut *15*, *30*
    clothing *246*, 247
shearwater, sooty 172
sheathbill *12*, 168, 195
Shemenski, Dr Ronald, medical evacuation 129
ships
    cruise 239, 298–9, 325–7
    research stations 239
Shirase, Lieutenant Nobu 11
sills 51–6, *52*, *53*
    Marie Byrd Land *57*
silverfish, Antarctic 166
Sirius Formation, *Nothofagus* fossil 64
Skottsberg, Carl, Swedish Antarctic Expedition *10*

skuas 176
Smith, David, sketch of huskies *19*
snow
    drifting 117
    on sea ice 69–74
snow flakes *77*
snowfall 78
soils 194, *195*
solar radiation 86, 103–5
    absorption by ozone 109–10
    atmospheric warming 109–10
    variation 155
solar wind, interaction with magnetic
    field *215*, 215
solar–terrestrial research 213, 217–21
South Georgia 5, *177*
    Discovery Investigations 13
    Sir Ernest Shackleton 14
    whaling 13, *14*, 177–8, 274, 276
South Magnetic Pole, French claim 6
South Orkney Islands *18*
    Argentine claims 17
    Dumont d'Urville 5
    maritime terrestrial ecosystem 193
South Polar Infrared Explorer
    (SPIREX) 226
South Pole, race to be first 12
South Pole Food Growth Chamber
    236, *237*
South Pole Station *see* Amundsen-
    Scott South Pole Station
South Pole Telescope 226
South Sandwich Islands *59*
    discovery 4–5
    formation 58–9
    maritime terrestrial ecosystem 193,
    194
South Scotia Ridge 138–9
South Shetland Islands
    early exploration 5
    maritime terrestrial ecosystem 193
    whaling 276
Southeast Indian Ridge 139
Southern Cross expedition 9–10
Southern Annular Mode 155–6
Southern Ocean 137–60, *138*
    biochemical properties 154
    climate change 303–4
    ecosystems 162–89
    extent 137
    flow beneath ice shelves 144–5, *146*
    fluctuations 154–6
    formation 139
    fronts 148
    observation 156–9

role in global climate 150–1
    swell 140
    warming 97
        and formation of sea ice 73
    waves *140*, 140
    winds *139*, 139–41
Southern Ocean Observing System
    323
Southwest Indian Ridge 139
space science research, future
    directions 227–8
space weather effects 213–21, *221*,
    222
sponges *179*
springtails 195–6, *197*
    radiation 202
stenothermy 312
Sterechinus neumayeri 180, 185
Stonington Island 17
storms *108*
    geomagnetic 215–17
stratosphere
    ozone absorption of UV radiation
        109–10
        ozone hole 129–32, 309–10
        polar stratospheric clouds 130, *131*,
        302
sub-Antarctic ecosystem 192–3
Subantarctic Front 146
Subantarctic Mode Water 146
subduction
    Drake Plate 42
    Gondwana 59–60, *60*
    Transantarctic Mountains 47
Sun, wave motion *224*, 224
SuperDARN radar 220
SWEAT hypothesis 46, *47*
Swedish Antarctic Expedition *10*, 10
swell, Southern Ocean 140

Tasmania-Antarctic Passage, opening
    139
temnospondyls 50
temperature
    atmospheric variation 109–10, *110*
    inversion 109, *113*, 113–15, *114*
    isotopes 81–4
    latitudinal variation 105, *106*
        influence of wind and currents
        105–6
    near-surface climate *111*, 111–15,
    113
    ocean water 141, *142*, *143*
    winter *115*, 115, 302, *303*
    see also global warming

Terra Nova Bay, polynya 122, *123*
terrestrial ecosystems 189–204
    biota 193–6
        history and colonisation 202–4,
        203
    climate change 314–15
    introduced species 204–5
Thaumeledone peninsulae 180
thecodonts 50
therapsids 50
Theron Mountains, sills 51–6, *52*, *53*
Thomson, Charles Wyville,
    oceanographic expedition 8
Thrinaxodon 50
till 98–9
Tomopteris 165
toothfish, Patagonian, illegal fishing
    297, *298*, 316
topography, influence on climate *112*,
    112–13
tourism 32, *33*, 318–19, 325–7
    airborne 325–6
    Antarctic Treaty 297–9, *299*
    cruise ships 239, 298–9, 325–7
    regulation 326
tracking, satellite, seabirds 170, *171*
Transantarctic Mountains 37–8, *39*
    barrier winds 120
    coal measures, Permian 48, *49*
    sills 51
    volcanism 61
Transit of Venus 1769 3, 7
transport
    research stations 237–43, *239*
        air 238–40, *239*
        land 239–43, *242*, *243*
        sea 239
    see also tourism
tropopause 107
troposphere 107
    temperature 109
tussac grass *190*
Twin Otter aircraft *38*, 129, 240,
    241

UAV (unmanned aerial vehicle),
    Terra Nova Bay polynya 122,
    123
United Nations, management of
    Antarctica 287–8
United Nations Intergovernmental
    Panel on Climate Change
    (IPCC) 132–5, *135*, *136*
United States of America
    claims on Antarctica 17

United States of America (cont.)
  Hughes doctrine 14–15
United States Antarctic Expeditions
  17
United States Antarctic Service
  Expedition 1939–41 17
United States Exploring Expedition
  5, 6
United States Navy, Operation
  Highjump 17, 277
Universe, observation 226
upwelling 147–8, 163
UV radiation 103, 109–10

Vaughan Williams, Ralph, *Sinfonia
  Antarctica* 32
Verne, Jules, Antarctic science fiction
  25
vertebrates, terrestrial ecosystem
  194–5
Victoria Land, Dry Valleys 192
  fossils *64*, 65
  lakes 200–1, *201*
volcanism
  aerosols 86
  Deception Island 42–3, *44*
  Gondwana 51–7, *57*
  Marie Byrd Land 60–1, *61*
  Mount Erebus 43–6, *45*
  South Sandwich Islands 58–9, *59*
  sub-glacial 46
  Transantarctic Mountains 61

waste
  field camps 242
  research stations *235*, 235
water
  isotopic composition, and
    temperature 81–4
  subglacial *99*, 99–100
water masses
  distribution 146–50
  formation 141
  frontal systems 147–8, *148*
waves, Southern Ocean *140*, 140
weather 102–36
  observation stations *118*, 118–19,
    119, 124
weather balloons *see* radiosondes
weather forecasting 124–9, *127*
  forecasters 128
Webster, William, Deception Island 5
Weddell, James, exploration 4
Weddell Abyssal Plain 138–9
Weddell Sea, Bottom Water 153
West Antarctic ice sheet 77–8, 80
  effect of global warming 99, 305–6,
    307
West Antarctic Rift System 60–1
West Antarctica *40*, 41
West Wind Drift 139–40, 149
Weyprecht, Karl 8
  International Polar Years 8
whales 174–6
  baleen, krill 174–5

blue 174–5
fin 174–5
first protection regulations 13
humpback *174*, 174–5
killer 174, 176
minke 174–6
right 174
sperm 174
whaling 175–6, 274, 312
  South Georgia 13, *14*, 177–8, 274,
    275, 276
Wilkes, Lieutenant Charles, US
  Exploring Expedition 5, *6*
Wilkins Ice Shelf, collapse 98
Wilson, Edward, Beardmore Glacier
  coal measures 48
wind 115–20
  barrier 116–20
  erosion *83*
  katabatic 115–16, *116*, *117*, 122,
    140–1
    in formation of sea ice 69
  role in energy budget 105–6
  Southern Ocean *139*, 139–41
winter, duration of *115*, 115
Winter Water 147
women in Antarctica *251*, 251
  hygiene 242
Wordie Ice Shelf, collapse 98
World Climate Research Programme
  (WCRP) 257
World War II 277